Airplane View of Greater Citadel—1933
(Photo by U. S. Army Air Corps, 2nd Photo Section, Langley Field, Virginia)

The Story of
THE CITADEL

By
COLONEL O. J. BOND

Southern Historical Press, Inc.
Greenville, South Carolina

This volume was reproduced from
An 1989 edition located in the
Publisher's private Library
Greenville, South Carolina

All rights reserved. No part of this publication may be reproduced,
stored in a retrieval system, transmitted in any form, posted
on to the web in any form or by any means without
the prior written permission of the publisher.

Please direct all correspondence and orders to:

www.southernhistoricalpress.com
or
SOUTHERN HISTORICAL PRESS, Inc.
PO Box 1267
375 West Broad Street
Greenville, SC   29601
southernhistoricalpress@gmail.com

Originally published: Richmond, VA, 1936
Copyright 1936 by: Garrett & Massie, Inc.
ISBN #0-89308-682-1
All rights Reserved.
Printed in the United States of America

FOREWORD

IN this book the writer has not had the purpose and makes no pretention of presenting to the public a history of The Citadel. In 1893, at the semi-centennial celebration of the institution, Colonel John P. Thomas, who was connected with it by the ties of graduate, professor, superintendent, member of the Board of Visitors, and historian, presented to the Association of Graduates his monumental *History of the South Carolina Military Academy*, covering the first half-century of its operation.

In 1942, some other alumnus may undertake the task of completing this history to cover the full hundred years of its work.

The author of this book has not undertaken to perform such an ambitious task, but only to tell the story of The Citadel in an informal way.

O. J. BOND.

The Citadel,
   August 18, 1932.

CONTENTS

PART I

CHAPTER		PAGE
I.	First Decade: 1822-1832	1
II.	Second Decade: 1832-1842	11
III.	Third Decade: 1842-1852	17
IV.	Fourth Decade: 1852-1862	40
V.	Fifth Decade: 1862-1872	54
VI.	Sixth Decade: 1872-1882	88

PART II

VII.	First Decade: 1882-1892	107
VIII.	Second Decade: 1892-1902	128
IX.	Third Decade: 1902-1912	156
X.	Fourth Decade: 1912-1922	178
XI.	Fifth Decade: 1922-1932	203
	Appendices	223
	Index	235

ILLUSTRATIONS

FACING PAGE

Airplane View of the Greater Citadel—1933	*Frontispiece*
Citadel Academy—1843 and 1850	7
The "Star of the West"	49
The Citadel—1865 to 1879	92
The Citadel in the Nineties	128
Colonel Oliver James Bond	166
Souvenir Picture of The Citadel—1910	172
Dress Parade at Old Citadel	183
Formation in Front of Old Sally-port	190
Alumni Hall	196
"Courtyard of a Thousand Arches"	201
Mary Bennett Murray Hospital	202
Barracks—Greater Citadel	210
Bond Hall—Main College Building	214

THE STORY OF THE CITADEL
PART I

The manuscript for the pages which follow occupied much of Colonel Bond's interest and attention during the closing years of his long connection with The Citadel. It is now published in memory of one whose life was devoted to the institution he loved, and whose influence will be felt throughout the years to come.

Mrs. O. J. Bond.

CHAPTER I
FIRST DECADE: 1822-1832

THE name Citadel indicates a military origin. The State's military college is, indeed, the outgrowth of an arsenal guard organized for the purpose of protection and not education. How the transformation came about is an interesting story which should be told to make the history of the institution complete.

On the site of the old Citadel, in the year 1822, there stood a peaceful building used by the State as a "tobacco inspection." It was then outside the city limits, just beyond the horn-work fortifications that guarded the town during the Revolution. A short street leading to it from King Street was known as Tobacco Street—a name long since lost to general public knowledge, but strangely preserved to this latter day in the street-lighting accounts of the Power Company. The changing of the "tobacco inspection" to the "Citadel" was due to a remarkable incident.

On Sunday, June 16, 1822, it is likely that the sun shone and sea breezes blew upon as peaceful a city as they do today, and the people went to their usual places of worship—to all appearances unaware that the menace of a frightful catastrophe hung over them. The daily papers had uttered no warning, and published only the usual unexciting news. But this was the day Denmark Vesey and his fellow negro conspirators had set for the massacre of the white population of the city.

The population of Charleston at this time numbered about 25,000—11,000 whites and 14,000 negro slaves and "free persons of color." Although one of the oldest of the American seaport towns, Charleston's growth in white population was remarkably slow, showing an average annual increase of less than a hundred during the period from 1790 to 1820. According to census figures, there was actually a decrease of nearly a thousand whites in the decade from 1810 to 1820. On the other hand, the colored population had been constantly increasing; the disparity in numbers between the two races was becoming a matter of serious concern.

In December, 1820, the legislature passed an act intended to abate this alarming condition. The preamble of this act recites "that the great and rapid increase of free negroes and mulattoes in this State, by migration and emancipation, renders it expedient and necessary for the Legislature to restrain the emancipation of slaves and to prevent free persons of color from entering into this State," and provisions were enacted "that no slave shall hereafter be emanicpated but by Act of the Legislature," and "it shall be unlawful for any free negro or mulatto to migrate into this State."

For some years, a spirit of unrest had been growing among the colored population, both slave and free. This was due to several causes. Certain local conditions, fomented, undoubtedly, by sinister antislavery propaganda from the outside, and the successful revolt of the negroes in San Domingo, furnished seed for revolution in Carolina. In 1816, an incipient insurrection of the slaves in Camden had been nipped in the bud.

The leader of the "intended insurrection" in Charleston in 1822 was in some respects a picturesque figure.

In 1781, in a cargo of about four hundred slaves imported by Captain Vesey into San Domingo for sale to the French, was a black boy about fourteen years of age who attracted the attention of the ship's officers by reason of his superior intelligence and fine physical development, so much so that Captain Vesey made a cabin boy of him, and retained him as his personal slave for many years, giving him the name of Telemaque, which in time became corrupted in the negro dialect into Denmark.

Captain Vesey, an old resident of Charleston, was probably often at home in his native city. At any rate, his negro slave, Denmark, drew a prize of fifteen hundred dollars in the East Bay Street Lottery in the year 1800, and used $600 of it to purchase his freedom from his master. He was then thirty-three years old.

Adopting the trade of a carpenter, Denmark Vesey came to be in the next twenty years a man of considerable influence among the colored people, due to his intelligence, physical prowess, and his bold and masterful manner. James Hamilton, intendant of Charleston, who wrote the official account of the "insurrection," thus describes him: "Among his color, he was impetuous and domineering in the extreme, qualifying him for the despotic rule of which he was ambitious. All his passions were ungovernable and savage; and to his numerous wives and children, he displayed the haughty and capricious cruelty of an Eastern Bashaw."

Vesey's activities were employed principally among the slaves. For several years, with the aid of a few trusted lieutenants, like Peter Poyas, Monday Gell, and the African necromancer, Gullah Jack, he preached discontent and hatred of the whites. An organization of considerable proportions was effected among the slaves in the city and surrounding country, and the utmost secrecy was enjoined regarding their plans.

This was the nefarious plot which was hatched: At an appointed hour, certain bands of negroes in the city were to seize the arms in the arsenal and guardhouse in the city and the arsenal on the Neck. The negroes from the islands were to come over and assemble in the lower part of the city, and similar groups from the country were to advance from the Neck. The combined forces would meet and proceed to slaughter the whites, achieve their freedom, and loot the town.

The zero hour was fixed for midnight, June 16, 1822.

THE STORY OF THE CITADEL

There was, however, one fatal defect in the plans of the conspirators. The slave trade had introduced representatives of many African tribes into America, some of the warlike and savage type of Denmark and Gullah Jack. But on the whole, the negro slaves in South Carolina were a docile, tractable, and well-disposed race of people, many of them devoted to their masters. It was inevitable that such a widespread conspiracy would become known to faithful slaves who would warn their masters of the impending danger.

Thus was the disaster averted. Forewarned, the city and State authorities combined to thwart completely the plans of the plotters.

On the evening of June 16, at 10 o'clock, the City Guard, the Charleston Riflemen, the Hussars, the Light Infantry, and the Neck Rangers were all under arms, and patrols and sentinels, with loaded muskets, were on duty throughout the city.

No doubt there was much peeping, whispering, and suppressed consternation everywhere among the negro quarters; but not a plotter appeared. It was a peaceful night. The record does not mention that a single shot was fired.

Next day, the civil authorities, acting with decision and despatch, proceeded to find and arrest the ringleaders.

The sessions of the court which was appointed to try them extended over a period of more than a month, and one hundred and thirty prisoners were tried. Of this number, thirty-five, including Denmark Vesey and his principal lieutenants, were convicted and executed. A number of others, less culpable, were transported, not to return on pain of death. About fifty were acquitted.

During the early conduct of the trial, it appears that there was some word of criticism made. Intendant James Hamilton, Jr., in his report on the case, closes with this statement:

"It is consoling to every individual who is proud of the character of his country, in the late unhappy events, to be able to say that within the limits of the City of Charleston, in a period of great and unprecedented excitement, the laws without even one violation, have ruled with uninterrupted sway—that no cruel, vindictive, or barbarous modes of punishment have been resorted to—that justice has been blended with an enlightened humanity, in according to those who had meted out for us murder, rapine, and conflagration in their most savage forms, trials which, for the wisdom, impartiality and moderation that governed them, are even superior to those which the ordinary modes of judicature would have afforded ourselves.

"With little to fear and nothing to reproach ourselves, we may, without shrinking, submit our conduct to the award of posterity, and ourselves to the protection of the Supreme Ruler of Events."

In order to prevent recurrence in the future of any similar insurrection, the legislature in December, 1822, passed "An Act to Establish a Competent Force to Act as a Municipal Guard for the Protection of the City of Charleston and its Vicinity." The Guard was not to exceed 150 men, and the area to be protected included two distinct districts, the city proper, which extended from the Battery to "the lines," and the Neck, which extended from "the lines" to the "cross-roads." The "lines" referred to were the fortifications built across the Neck in the War of 1812—practically coincident with what is now Line Street. The "cross-roads" referred to the intersection of the Bee's Ferry with the Meeting Street Road, about five miles further up the Neck.

The act provided first "that the land and buildings now used as the tobacco inspection be invested in a board for the purpose of being fortified as an arsenal and guardhouse for the municipal guard," and second, that certain "Lands on Charleston Neck be sold by the board and the proceeds used for the erection of suitable buildings." The "fortified arsenal" on the site of the "tobacco inspection" ultimately became known as the Citadel, and the buildings on the Neck as the Magazine.

In *Mill's Statistics*, published in 1826, an account of the buildings at the Magazine is given with such particularity of information that it would seem likely that the famous architect was himself the designer:

"The State is now erecting powder magazines upon a new and permanent place which enables these buildings to be placed with safety much nearer to the built part of the city than they have hitherto been; more accessible to our citizens, and more under the protection of the Neck guard.

"These magazines are distant about two miles four furlongs N.W. of the courthouse, situated on an island formed by a creek making up from the Cooper River, and navigable at any time of the tide to the very spot. The powder magazines are nine in number, all of circular form, with conical roofs, and disposed in three ranges 130 feet apart. The center building is the largest, and intended exclusively for the public powder. It will contain, upon an emergency, four thousand kegs. The roof is made bomb-proof. The surrounding buildings are made large enough to contain each one thousand kegs, though it is never intended (except in case of necessity) that more than half of this quantity should be deposited there at one time."

In the year 1930, only some dilapidated old ruins of the magazine remained. Laurel Island, on which it was situated, is now undistinguishable from the mainland, and is half a mile inland from the Cooper River, the extensive marshland having been filled in by dredging from the river. The flow of silt extended into the old Magazine property, covering up at one place under the spreading live oaks the tombstones of several 'soldiers of the Citadel,' who had died of yellow fever in the thirties.

1   1   1   1   1

The "tobacco inspection" had been connected with an evanescent phase of Carolina agriculture. A very interesting portion of the history of the State is concerned with the various crops which have been the staple products of the farms at various times. The cultivation of rice, for instance, for which South Carolina was early and for long noted, had practically disappeared by 1920. Governor Glen, writing in 1754, said: "Our chief and almost our sole dependence hitherto has been on rice, but we now have happily found out other resources; we no longer plant indigo by way of experiment."

Statistics for the year 1748 show that 55,000 barrels of rice of a value of £618,750, South Carolina currency, were exported from Charleston. Indigo had then made a good beginning as a rival of rice—134,118 pounds, of a value of £117,353, South Carolina currency, having been exported in that year. Indigo, however, ran its course as an agricultural crop long before rice began to decline. Today, it is only those farmers versed in the history of the State that know it was ever cultivated. It is interesting to note, in view of subsequent events, that the export of "cotton wool" in the year 1748 was a meager seven bags, of a value of £175. In the same year, no mention is made of tobacco as a product for export. Sometime before the Revolution, however, it began to be cultivated, and by 1784 was an important crop, although its curing and marketing presented serious problems that were not satisfactorily solved for many years.[1] In 1784, the legislature passed "An Act to Regulate the Inspection and Exportation of Tobacco," on the grounds, frankly stated in the preamble that it was "necessary to inspect the article of Tobacco before the same is exported to foreign markets, or consumed at home, in order to prevent its being brought into discredit by fraud or negligence of those who shall cultivate or export the same." The establishment of "tobacco inspections" was the result.

Official warehouses, under the supervision of duly appointed commissioners, were established in various parts of the State. The one in Charleston was first operated on Gadsden's Wharf, but about the year 1790 a building of some pretentions was constructed for the purpose on the plot of ground which afterwards became known as Citadel Square. This area of about ten acres had been ceded by the State to the city of Charleston

---

[1] *The South Carolina Gazette*, for March 23, 1795, contains an advertisement of land for sale, adapted for the cultivation of "rice, corn, indigo, or tobacco."

when it was incorporated in 1783; a strip about 150 feet wide along the northern boundary of this square was bought back by the State March 13, 1789, for the site of the "tobacco inspection." The rest of the square was retained by the city for a muster ground and other purposes. Fraser, in his *Reminiscences,* mentions that on February 22, 1797—Washington's birthday—"the two regiments of the City, together with the battalion of artillery, had been reviewed by Gen. Washington near the Tobacco Inspection."

About the beginning of the nineteenth century one of those radical changes in agriculture which have characterized the history of the State began.

The accession of King Cotton in the agricultural life of the South was due to a sickly young man from Massachusetts who came South in 1793 and lived for a while in the home of Mrs. Greene, widow of General Nathaniel Greene, in Savannah, Georgia. This young man had a noticeable, knack of making mechanical contrivances, a talent which led to a casual commission to devise a machine for removing the seed from cotton. Up to this time, this work had been done laboriously by hand, it being considered a fair day's labor for a slave to separate a single pound of fiber from the seed. The young man at this time had never seen a lock of cotton. But with Yankee confidence, he undertook the commission, and thereby wrote large the name of Eli Whitney among the notable inventors of all time.

It may be said, in passing, that while it would be difficult to estimate the influence of the cotton gin upon the South, the country, and even the world, the invention—by a not uncommon trick of fate—brought to its creator more trouble than reward.

As a result of Whitney's epoch-making invention, the production of cotton in the Southern states grew with astonishing rapidity. It is related that in 1784 eight bags of the staple were seized at the customhouse in England as fraudulently entered—on the ground that "cotton was not a product of the United States." In 1821, the production in the Southern states amounted to 250,000 (modern 500-lb.) bales, which increased in 1832 to about 650,000 bales, a large part of which was exported and manufactured in English mills.

During the second war with Great Britain, Charleston's commerce had been nearly destroyed by the blockading British fleet; but with the return of peace and the increasing production of cotton, the city began to see prosperous days again. Fraser says: "Cotton had been found to pay the planter so much better than tobacco that in a few years it entirely superseded it. The inspection buildings, put up at so much expense, were taken

Citadel Academy — 1843

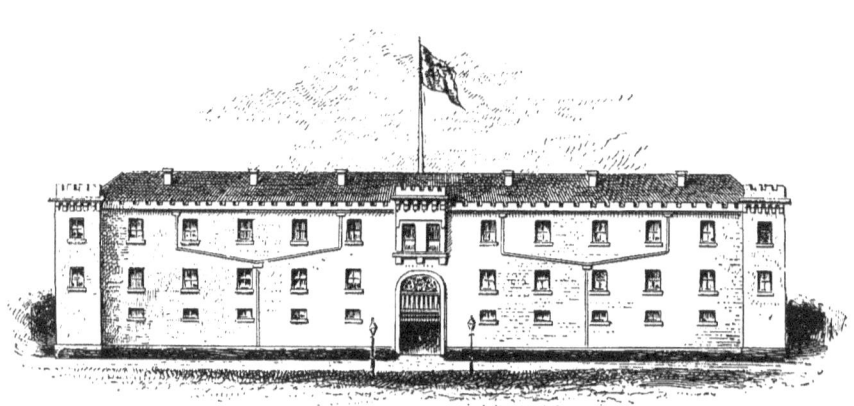

Citadel Academy — 1850
(Pictures by courtesy of Walker, Evans & Cogswell Company)

down, and the multiplied enactments regulating the sale of that staple became a dead letter."

When the State, therefore, in 1822, was in need of a site for an arsenal, the tobacco inspection was ready at hand.

The commissioners charged with the construction of the building engaged Frederick Wesner as architect. It appears that for some reason the work did not progress with much expedition, and it was not until the close of the year 1829—seven years after the passage of the "Act to Establish a Competent Force to act as a Municipal Guard for the Protection of the City of Charleston and its Vicinity"—that the *"State* arsenal"[2] or Citadel, was ready for occupation.

In the meantime, the State and city authorities had changed their plan for a municipal guard, and joined in a request to the War Department that United States troops from Fort Moultrie be detailed for the purpose.[3] This request was complied with, and the occupation of the new building by these troops was made the occasion of a patriotic celebration in conjunction with the observance of the fifteenth anniversary of the Battle of New Orleans, January 8, 1830. An interesting account of these exercises is given in *The Mercury* of January 11, 1830:

"The State Arsenal being ready for the reception of the regular troops destined to occupy it, Friday last, the 8th of January, was selected as a day suitable to the ceremony of delivering it into their possession.

"Ten volunteer companies of the Sixteenth and Seventeenth regiments were detailed as an escort to this guard, whilst the Washington Light Infantry were dispatched to take possession of the Citadel as the guard on duty to be relieved. About 9 o'clock the steamboat *John David Mongin* arrived at Fitzsimmons' wharf with three companies under the command of Major Heileman, and from thence marched to Broad street, where they were formed in line in front of the Custom House. The escort was formed in Broad street, the right on Meeting street. About 11 o'clock the escort under the command of Captain Egleston moved down Broad street, and upon Major Heileman being apprized that all was ready, the line of march was taken up, and upon the military passing the city hall, in which the intendant and the council, the commissioners who had superintended the erection of the building, the officers of the United States army, foreign consuls and distinguished strangers had previously assembled, they formed

---

[2] So called to distinguish it from the *United States* arsenal in another part of the city, now the barracks of the Porter Military Academy.

[3] Resolution of the South Carolina legislature, December 17, 1829: *"Resolved also, that the intendant of the City of Charleston be and he is hereby authorized and directed to occupy the state arsenal by a detachment of United States troops conformably to an arrangement already made for that purpose with the Department of War of the United States."*

in the rear of the military, followed by a large concourse of citizens. The procession moved up Meeting street, and upon arriving near to the arsenal, the military formed into line with presented arms, whilst the civil procession moved through the great entrance into the interior of the arsenal and took possession of the immense balconies that nearly surround it within. The United States troops then passed the state troops, and that portion intended to occupy the station entered and took post on the right of the Washington Light Infantry. The remainder, together with the state troops then entered and formed a line around the square.

"The ceremony of relief then took place, Col. Hunt acting as old officer of the day. The new guard having passed inspection, they were marched past Major Heileman, acting as the new officer of the day, were received with military honors, and took position to the right of the old guard. The ceremony of relief having concluded, and the arsenal having been delivered to the charge of Major Belton, commanding the United States troops, the officers of the United States army were invited to the saloon and introduced to Col. Hunt, severally, to the Intendant and Members of Council who had previously been stationed there. After the ceremony of introduction, the Intendant addressed Major Heileman and his brother officers, explaining the object of the Institution, and the duties they were expected to perform; reminding them of the order of the War Department; assuring them of the full reliance reposed in them both as soldiers and friends; and welcoming them cordially in the name of the community. He also made some brief remarks in allusion to the day (the 8th of January), which had been selected to receive them; upon the happy illustration which their establishment amongst us afforded of the true principles of our government, of the confidence with which a State may demand assistance, and of the readiness of the national administration to afford it. He concluded by inviting Major Heileman and his brother officers, and all the company present to a collation prepared for the occasion. The collation was very numerously attended, and presented an animated scene of republican festivity.

"The large and splendid Area, lined as it was with the Military in elegant uniform, overlooked by the immense balconies thronged with the beauty and delicacy of our fair country-women gave to the scene an inspiring effect, that can not be realized by any that did not partake the gratification which the ceremonies offered. The United States troops were all well dressed and made a brilliant display. Our volunteer Corps were unusually numerous, and in dress and appearance exhibited more than usual elegance. The weather was remarkably mild, the sun shone with splendor, the skies were divested of clouds, and the surface of the earth was just in a condition to render walking pleasant. Upon the whole, the associations connected with the day, in conjunction with the brilliant and imposing ceremonies performed, could not fail to produce in the heart of every patriot sensations which none but a free born American citizen can enjoy."

This incident, involving apparently the cordial coöperation of the national, State, and city authorities, is the more remarkable because it occurred on the very eve of a controversy which was not only to divide the people of the State into bitterly antagonistic factions, but was to array South Carolina against the Federal Government in a struggle which would ultimately shake the Union to its foundations.

And if it be asked: "What was it that caused all this dire delirium?" the answer that would be most nearly accurate, if it had to be stated in a single phrase, would be: "The invention of the protective tariff."

At the beginning of the year 1830, an antinationalist sentiment had already perceptibly developed in the State, fostered chiefly by antagonism to the Tariff of Abominations of 1828. From condemnation of the tariff, which was practically universal in the agricultural South, to opposition to the Government was an easy step. How far would this opposition go? Nullification was the remedy evolved by South Carolina; but it did not require the gift of clairvoyance to perceive the spectre of Disunion lurking in the background.

Sentiment in the State was by no means unanimous. One faction of the party was willing to confine the fight on the tariff to the halls of Congress and the decisions of the Supreme Court. Another group, which rapidly grew in numbers and power, favored State action, even if it led to secession. The discussion waxed warm, and in the five-year period from 1830 to 1835 the whole attention of the people seemed turned from peaceful pursuits to the political forum, whereon the foremost men of the State debated State's Rights and the Federal Constitution.

The climax came on November 24, 1832, with the passage of the historic "Ordinance of Nullification."

This was followed by a proclamation of the President of the United States—the only president born in the State of South Carolina—advising his "fellow countrymen" that the tariff laws would be enforced. There were reports that Old Hickory talked of sending armies and hanging traitors. Undoubtedly, war clouds were on the horizon.

Governor Hayne was not neglectful of military preparations himself. Responsible citizens in all parts of the State were busy organizing companies which would fight for the State if necessary. In February, 1833, James H. Hammond, military organizer in Barnwell district, advised the governor that over nine hundred men in Barnwell—about two-thirds of the male population of military age—had volunteered for service.

Incidentally, a resolution was approved by the General Assembly on December 10, 1832, requesting the governor "to intimate to the commanding officer of the United States troops now in garrison at the State Citadel in Charleston that he make an arrangement to remove them as early as possible, the accommodations of the building being needed for the arms of the State." The request was complied with, and the *City Gazette* of December 25, 1832, published the brief news item: "*Military.* The company of United States troops who have so long garrisoned The Citadel, by the special request of the State and City authorities, evacuated that post yesterday, and proceeded to Fort Moultrie."

Thus ended the first occupation of the Citadel by United States troops—January 8, 1830, to December 24, 1832. Many years afterward, on February 18, 1865, the second occupation of the Citadel occurred, when a detachment of the Federal Army took possession of the buildings, this time without invitation or permission, and held them for a period of seventeen years.

CHAPTER II

SECOND DECADE: 1832-1842

WHEN the Citadel was vacated by the United States troops on Christmas Eve, 1832, the governor ordered the Magazine Guard on Charleston Neck to occupy it. This guard was augmented to sixty men under the command of Captain Charles Parker and Lieutenants Edward C. Peronneau and William S. Gaillard. From the Citadel a daily guard was detached to the Magazine.

In the brief space of three years, the military significance of the Citadel had undergone a startling transformation. Very few persons, seeing the extensive warlike preparations going on there in January, 1833, would have recalled that it was established to act as a municipal guard for the protection of the city of Charleston and its vicinity against the recurrence of a negro insurrection. The stupendous fact was that South Carolina was preparing to defend herself against the United States of America!

Fortunately, the struggle was averted. The great pacificator, Clay, brought about a compromise measure in Congress which was accepted by both sides as a temporary solution of the tariff question. The ordinance of nullification was rescinded, and both Unionists and Nullifiers celebrated the peace with elaborate entertainments in Charleston. The latter gave a grand military ball in the quadrangle of the Citadel on March 27, 1833, which was facetiously described by a gentleman from Maine, residing temporarily in Charleston, in a letter to his home paper, and quoted at length in Theodore D. Jervey's *Robert Y. Hayne and His Times*. Evidently our Yankee visitor did not take the nullification controversy as seriously as the South Carolina statesmen:

"The Nullifiers are doing things in grand style. This Charleston is no laggard in working off a fête. The Nullifiers are men of taste, men of little guns and men of big guns, swords and cutlasses, great spunk and fine speeches, pretty ladies and pretty dances. Who would not be a Nullifier and live in such a land, feed on such chivalry and enjoy such a ball? . . . As a Yankee under good auspices I went last evening into the Citadel, the heart of the Nullifiers' camp, and among big-mouthed cannon, muskets, fusees, pistols, long swords and short swords, King's arms, rifles and fowling pieces, spears, pikes and bayonets bristling for horrid war, I found—think what? Not less than 1,200 ladies. What a place to put ladies in, good-hearted creatures, if they are like our Northern belles and fair ones. . . . Well I went to the ball at 8 o'clock or a little before. It was in the Citadel, which is the armory of the state, where are deposited Carolina's munitions of war, with which she was going to whip her twenty-three sovereign sisters, with men enough to eat her up, slaves and all, if they gave the Kentuckians but the quantum of an eye and ear apiece. The Citadel is an oblong building, perhaps two hundred feet in length, and with an open area in the center, perhaps sixty feet in length. This area was floored over for the occasion, a canopy overhanging, and thus a grand, magnificent hall was prepared. The armories answered for drawing rooms. . . . We hung our hats on bayonets. Their muzzles answered for candelabras. Around the outside door was a vast multitude of white people, black people, and yellow people, with not a few nondescripts. Pillars and arches of lights of almost all colors formed by variegated glasses, in which were the lamps, immense in numbers, were thrown around the door. . . . 'Nullification is the rightful remedy,' quoted from Jefferson in large capitals, glared the spectator in the face. Rockets and bombs were let off in all directions. From half past seven until nine, carriages in line were discharging men in epaulettes, plumes, palmetto buttons, green coats, gray coats, red buttons, white breeches, yellow breeches, and black breeches. All the soldiers, the volunteers of the Empire, came in the uniform of their corps. And carriages were discharging ladies also, two at least to each gentleman. Ladies in white, in black, in scarlet, in blue; ladies in hats and feathers of all fashions. No two ladies looked alike . . . Now let us go into the hall. A more magnificent picture was to be seen. We ascended a flight of stone stairs, walked along an ornamental piazza or corridor interwoven with imitation flags of cambric muslin of red and white and sprigs of cedar, live oak, and palmetto leaves. Ranges of card tables were spread in the gentlemen's drawing rooms. Rivers of wine were near. Refreshments of ices, lemonade, etc. One's head and hair adjusted and hat disposed of, he was ushered along the gallery, so as to view the company below, who, now the governor had entered in uniform and epaulettes, and General Hamilton also, in all the pomp of the camp, with their respective suites, prepared to dance. Cotillions were formed in the crowd, with exceeding difficulty; but when they were formed, the black band, who were planted somewhere on high, began to sound with horn and clarinet and drum and cymbal, and I know not what other instrument, but they made a deafening noise. I took the opportunity to go below, to run among the groups, to see the cannons, etc. . . . Under the staging for the band were long iron pieces of ordnance, with their mouths turned to the company. Back of them were five ranges of supper tables. . . . Between the columns were medallions with emblematic devices on which were compliments to distinguished Nullifiers in South Carolina. Calhoun had one, and was called 'the great luminary'; McDuffie had one, and was said to have the eloquence of Henry and heart of Hampden. Hayne had one with an extract from one of his speeches. Hamilton had one, which I have forgotten. W. R. Davis and Barnwell had only one, which was not fair, for why should they not have had one apiece? Pinckney had one, Sumter had one, and was called 'an old cock, whose last crow was for liberty'; Trumbull had one, which called him Brutus. . . . Enjoying all this, and thus in the heart of the Nullifiers' camp, I ran around some gentlemen and ladies with that perfect independence in which obscurity always clothes one. I knew but few, and could not find that few very often in the multitude. Here was a bevy of ladies discussing the merits of the Yankees and Yankee women. There is a platoon sweeping and demolishing a half-formed cotillion. Here is the Governor of the State in cap and plume and epaulettes, with his amiable lady, wearing the cockade of Carolina. There ex-Governor Hamilton, Emperor of the South, far less humble than Napoleon when only trampling on the throne of Europe. Here was a cluster of Generals, Colonels, and captains, epauletted to the ear, with swords dangling between their feet, with spurs sticking in their heels. There a body of men vaunting the prowess of Carolina. Carolina! Carolina! Who will not stand for Carolina? . . . The Hayes, the Hamiltons, the Sumters, the Pinckneys, the Calhouns, the McDuffies, the Millers, the Turnbulls of Carolina. Huzzah for Carolina! . . . Talking of Nullification dying out, it is nonsense, when you work upon the passions and the feelings of people with such shows. Every man and child there will live and die a Nullifier. I was half a mind to become one myself—splendid, mad people, if this meets your eye, this letter from not an ill-natured spy in your camp, pray take his advice and get sober again. Leave off drinking these intoxicating draughts of Carolina chivalry. . . . Ladies, don't hate the Yankees, the d--n Yankees, as some of your beaux term them. Upon my word, we are not all tin peddlers, not all hucksters, wooden nutmeg, wooden ham sellers, though we live in such a cold, rocky land, we must depend in part on our wits. Some of us are honest and won't cheat you. . . . Come down among us and you will find that we are not all icicles or fog-banks. We like you better than you like us, and speak better of you, though you have two faults to our one. . . . We go for the Union, because duty, patriotism, and common glory look that way, not because we are more interested in it than you. Hoist up the Star-Spangled Banner in your Citadel. Let us all be Americans, all Carolinians, all Yankees."

The relief felt at the adjustment of the Nullification issue did not deceive the leaders of the two factions into the belief that the matter was permanently settled. Such men as Hayne, Hamilton, and McDuffie saw clearly that they were at the beginning rather than at the end of the contest. They were impressed, in consequence, with the importance of military preparedness on the part of the State. What Hammond had written the governor from Barnwell in February, 1833, was true in many other parts of the State: "every one seemed ready to fight; the difficulty with us will not be the want of men but *officers and means*. . . . We should by all means have a military department in the college."

In 1833, the legislature passed an act "more effectually to provide for the defense of the State," appropriating a large sum for arms and munitions, and establishing two principal arsenals and magazines, one at Charleston and the other at Columbia. The smaller arsenals scattered over the State were then to be abolished. The Citadel, in the city of Charleston, and the State Magazine on Charleston Neck, were already serving this purpose; but it was not until five years later, 1838, that the arsenal and magazines at the capital were completed and occupied. The site selected for these was an eminence in the town which has since been known as Arsenal Hill. The only building of this group which has survived to this day is the present governor's mansion, the rest having been destroyed in the burning of Columbia in 1865 by Sherman's army.

The importance of military education was also emphasized by Governor McDuffie in his messages to the General Assembly. In 1835, he called to the attention of the legislators "the expediency of combining in our general system of school instruction the use of arms and the elements of military tactics with the common branches of education." Again, in 1836, he suggests "that the young men of the (South Carolina) College be organized into one or two corps of cadets, by law or by the regulations of the institution, and be required to devote certain hours to the exercise of drilling under the superintendence of the military professor." These recommendations, however, bore no immediate fruit.[1]

---

[1] In the same message, the Governor says: "In one of the most distinguished Grammar Schools of the State, a company of Cadets was formed almost under my own eye, and while their improvement in tactics was striking to every observer, the intelligent gentleman at the head of the institution assured me that he derived great advantage in its government from the manliness and sense of honor imparted to the young men by the change in their mode of recreation." What school was it?

Dr. D. D. Wallace, author of *The History of South Carolina,* is disposed to think that this reference may have been to the Rice Creek Springs school in Richland district, which was situated on the Columbia-Camden road, about fifteen miles from the capital. He cites an article from the Charleston *Courier* which was copied by the Camden *Commercial Courier* of December 9, 1837, saying that "South Carolinians like military education"; that "they support Capt. Partridge's school and Rice Creek Springs" (The first reference was no doubt to the Military Academy founded by Captain Alden Partridge at Norwich, Vermont, in 1819). The editor of the Camden *Journal,* at an earlier date (August 1, 1829), speaks of the Rice Creek Springs school in glowing terms, as the result of a personal visit. He was much impressed with the fact that the student received so much personal instruction, there being as a rule one teacher to twelve pupils. Some of the teachers had been trained in the best Northern universities.

About this time, an experiment was tried in Virginia which attracted the favorable attention of Wm. C. Preston and other statesmen in South Carolina. A State arsenal had been established in Virginia at the town of Lexington in "the Valley" about the year 1816, where were stored about 30,000 stands of arms and other military matériel. It was guarded by a company of twenty-eight men. It is probable that at times the conduct of the soldiers in the little town met the disapproval of the citizens. At any rate, the suggestion of establishing a school at the arsenal, in which students would be substituted for the soldiers, was debated as early as the year 1834. In the following year, several articles over the pen name Civis appeared in the public press favoring the project, and made so favorable an impression that in 1836 an act was passed to carry it into effect. At first it was proposed to make the school a part of Washington College (afterward Washington and Lee University), whose campus adjoined the arsenal grounds. Later, however, due principally to the influence of Colonel J. T. L. Preston, of Lexington, who was revealed as the author of the Civis articles, an independent school was organized under the title of the Virginia Military Institute, and on November 11, 1839, this historic institution, which was to earn a distinguished reputation in the nation as well as the State of Virginia, opened its sally port to the first corps of cadets—thirty-two young Virginians.

With the Virginia example in view, during the administration of Governor J. P. Richardson, and no doubt at his suggestion, a bill was introduced into the South Carolina Legislature in 1841 by Colonel John Phillips, of Charleston, to convert the Arsenal at Columbia into a military school. The act proposed that the cadets at the Arsenal should have the use of the library at the South Carolina College, and get the benefit of such scientific courses there as its faculty would permit.

Although this act failed of passage, the governor was not at all discouraged, and thought it within his discretion to make a trial of his scheme under the existing law. He accordingly dismissed the soldiers at the Arsenal and enlisted in their stead a number of boys from the several districts of the State selected by the Commissioners of the Poor. A system of instruction was instituted which included not only the necessary military duties of the guard, but also some of the ordinary educational courses. In his message to the legislature next year, he refers to "the very satisfactory

---

An advertisement of this school in the Camden *Journal* of January 3, 1829, emphasizes this feature, stating that "the boys live twelve to a cottage, each with its teacher, who joins in their recreation as well as studies." A vacation of three weeks a year only was allowed. The subjects taught included belles-lettres, commerce, agriculture, and both military and physical education. The rates were $250 to $275 a year. The superintendents were R. W. Bailey and H. L. Dana.

success of the short and limited experiment which it was within my official discretion to institute."

In the summer of 1842, the governor dispatched Major Charles Parker, of the Citadel, to Lexington to examine into the work and character of the Virginia institution. The governor also visited his friend, General James Jones, a former adjutant general, at his home in Vaucluse, and enlisted his interest in the proposal to convert both the Citadel and Arsenal into military schools.

On the editorial page of *The Courier* of November 22, 1842, a letter from Civis *(another* Civis, probably), cited the example of the flourishing Virginia Military Institute in eulogistic terms, and closed with the suggestion: "As this institution has so far more than realized the expectations of its friends without additional expense to the State, it may be well worth the attention of the Legislature to consider how far the example of Virginia is worthy of imitation."

When the legislature convened in the last days of November, 1842, Governor Richardson was ready with a strong message favorable to the project. He presented a convincing argument on the advantages of combining the military duties required of the guards at the Citadel and at the Arsenal with a system of education for the poor but deserving boys of the State. The conduct of the soldiers of the Arsenal had not always been exemplary; and it appears that the system of public instruction for "the indigent poor," for which the State had made an appropriation of $36,000, had proved unsatisfactory, for the governor says: "The unprofitable use of the annual appropriations of the State to establish a system of public instruction constitutes another strong inducement to prosecute an experiment which promises by its fruits to form *one* exception, at least, to the hitherto entire and unmitigated failure of all her efforts to educate her indigent youth." In advocating his scheme to found military academies at the State arsenals, he concludes as follows: "If the success of these institutions should form the basis of future and important improvements which may judiciously be extended to our free schools; if they should supply better teachers from their alumni; if they should suggest higher standards and better systems of morals and tuition; or if they only awaken greater ardor in the people, and a warmer interest in our rulers to advance the cause of education, they will achieve more for the weal and honor of our State than all the other labors and appliances of government could in any other manner confer."

CHAPTER III

THIRD DECADE: 1842-1852

A BILL embodying the governor's ideas was introduced in the House by his friend, General D. F. Jamison, of Orangeburg, chairman of the military committee. It met with general approval—only one member, Ashby, newly elected, seizing the opportunity to win his spurs in debate by opposing the measure. This early pacifist advocated the abolition of the Citadel's guard and turning over the building to the Charleston, Louisville, and Cincinnati Railroad for use as a "deposit" for merchandise.

Governor Richardson's proposal, however, had won the common sense views of the legislators, and the bill was passed with little opposition on December 20, 1842—a date forever memorable in the history of the Citadel as the birthday of the institution.

The act empowered the governor to appoint a Board of Visitors consisting of five members, and the Adjutant General, ex-officio. (The next year, 1843, the governor, ex-officio, was added to the Board.)

Governor Richardson having retired from office on December 7, it devolved upon his successor, Governor James H. Hammond, to appoint the first Board of Visitors, which he named on the day after the passage of the act, December 21, and which consisted of General James Jones, who was chosen chairman; D. F. Jamison, W. J. Hanna, Daniel Wallace, and J. H. Means. The ex-officio member was General James W. Cantey.

On the same day the members, with the exception of Messrs. Wallace and Means, met at the arsenal in Columbia to discuss plans for the organization of the schools. The organization of the Arsenal Academy was the first object of the Board's concern, and the election of a superintendent for this academy was fixed for January 11, 1843, an advertisement being inserted in the newspapers to this effect in order to give candidates for the position an opportunity to apply to the Board.

At this first meeting, also, the important task of preparing regulations for the organization and government of the academies was entrusted to the chairman. In this work, General Jones undoubtedly followed the military academy at West Point in its regulations for the practical military instruction and the discipline of the cadets. In the curriculum of studies, however, he did not have to follow the West Point course so closely, since the students after graduation were not to enter the military service, but would be absorbed into the general citizenship of the State.

It may well be that General Jones also took a suggestion from Norwich

University, in the State of Vermont. This institution had been founded in 1819 at the town of Norwich by Captain Alden Partridge, a retired Army engineer officer and a graduate of West Point. He had been an instructor on the faculty of the Military Academy for ten years and was the superintendent from 1815 to 1817. He resigned from the Army to found a school to carry out his own ideas of what education should give. His school was called at first by the comprehensive but cumbersome title of The American Literary, Scientific, and Military Academy, but was later (1834) changed to the present shorter title.

Captain Partridge was a man of originality. His design was to found an institution which should combine the valuable discipline of the military school with the more liberal course of studies of the liberal arts college. The principal features which he emphasized were: practical courses adapted to instruct young men in the various duties which an American citizen may be called on to discharge; attention to physical education; elimination of too much idle time for students; and the regulation of their spending money in order to prevent extravagance and dissipation.

He believed in military education, but not in a large standing army. He says: "Let practical and scientific instruction be a part of our system of education, and we shall become a nation of citizen soldiers; the need of a large standing army will be done away with; in case of sedition or foreign invasion, a sufficient force will be ready to take the field, and when the emergency passes away the character of the soldier will be lost in that of the citizen."

It is remarkable that in the two decades from 1820 to 1840 about one hundred South Carolina boys travelled to this distant Vermont school; and it is quite probable that General Jones knew some of these young men, and something about the institution which they attended. At any rate, the seed which the Board of Visitors planted developed into a military school very similar to Norwich University and the Virginia Military Institute.

At the second meeting of the Board held at the Arsenal on January 12, the officers of the Arsenal Academy were elected. These were: Alfred Herbert, Superintendent; Joseph Matthews, Professor and Bursar; and Dr. A. H. Nagel, Surgeon.

Captain M. C. Shaffer, who had been in charge of the Arsenal Guard, was retained as powder receiver until October following.

The election of the Citadel officers was deferred until a month later, February 24, when the following were selected: W. F. Graham, Superintendent; F. W. Capers, First Professor; J. E. B. Finley, Second Professor; Dr. H. Boyleston, Surgeon; J. Ladson Gregorie, Bursar; and William Yeadon, Arsenal Keeper.

At this meeting, eight beneficiary cadets and six pay cadets were appointed to the Arsenal, and twenty beneficiaries to the Citadel, these to report to their respective schools by March 20. The roster of these first appointments to the military academies includes:

To the Arsenal: James McCan, R. D. Bevill, Daniel McKagen, McBelton O'Nealle, C. G. Billings, Joseph Howell, Ralph Burnet, and Luther Williams, beneficiaries; and Zachariah Cantey, Thomas H. Ligon, Thomas J. Summer, G. W. Lamar, James W. D'Lyon, and Bennet Crafton, pay cadets. To the Citadel: William Staley, Chas. C. Tew, Robert J. Lester, Washington Cross, Robt. D. Landreth, C. Stopplebein, Judah Alexander, Thomas F. Salinas, Chas. O. Lamotte, Wm. L. Lockwood, Theodore Plane, Wm. W. Wilbur, Jno. Campbell, Robert Simons, Alex P. Brock, Samuel Dixon, Wilbrandt Schmidt, Henry Caskin, John H. Swift, and Jos. P. Reeves. Eight additional beneficiaries were appointed to the Citadel on May 3: Jos. H. Buckner, John Branch, Edward Kingsmore, W. R. McDonald, John Bowen, J. W. North, Richard Hay, and W. J. Magill.

On this date, also, the first cadet from outside the State, Henry Slappey, of Georgia, was admitted as a pay cadet at the Arsenal.

Thus, about March 20, 1843, the Citadel and Arsenal academies began their sessions.

In December, 1843, the Board made its first report to the legislature. On the whole, the first year's work of the academies had been satisfactory. To quote:

"In arranging the course of studies for the academies, the Board have aimed at a system of education at once scientific and practical, and which, if its original design is carried out, will eminently qualify the cadets there taught for almost any station or condition of life. During the course, besides the usual branches taught at the primary schools in the State, they will be instructed in the history of South Carolina, in modern history, the French language, every department of mathematics, bookkeeping, rhetoric, moral philosophy, architectural and topographic drawing, natural philosophy, chemistry, geology, mineralogy, botany, civil and military engineering, the Constitutional Law of the United States, and the law of nations. In addition to that course, they will be instructed in the duties of the soldier, the schools of the company and of the battalion, the science of war, the evolutions of the line, and the duties of commissioned officers.

"Of the above mentioned course, only the most elementary part, comprising those of geography, arithmetic, English grammar, history of South Carolina, and duties of the soldier, have been taught. No cadet who applied during the past year was found sufficiently advanced to be placed in any beyond the fourth, or lowest, class; and indeed, in very many instances their previous course of instruction had been so imperfect that the professors found themselves compelled to instruct them in spelling, reading, and writing, before they could proceed to the course of studies laid down by the Board."

With the military features of the schools the Board was greatly pleased:

"In their military training, they exhibited a manly and soldier-like bearing, and the Board found it difficult to realize the fact that such a change had taken place in the appearance and conduct of the boys who, less than a twelve-month ago, came into those institutions careless of their persons, awkward, and untaught. Besides affording sufficient protection to the arms and public property at the two posts (far greater than the old Guard which they displaced), their military training facilitates their instruction in the branches of study by habits of good order and discipline which it promotes, and in the opinion of the Board appears to solve the difficult problem of the management of a number of young men in institutions of learning and science. By requiring them to account for every moment of their time, it prevents them from acquiring vicious habits, and withdraws them from the allurements of dissipation. Nor does the discipline to which they are subjected appear to weary them; on the contrary, they seem to be satisfied and happy; and the Board believes that a dismissal from the Academies would be regarded by any of the boys as a severe punishment.

"The Board assures your Excellency that this first year's experiment has succeeded beyond their hope; and they confidently predict that its future success will not disappoint the Legislature in changing the nature of these institutions."

The act founding the schools had given to the Board of Visitors large powers and wide discretion. They were to shape the destiny of the institutions throughout the years. The historian of the Citadel pays this tribute to them:

"Fortunately for the new Military Schools of the State, now put in operation, the Board of Visitors consisted of wise men, devoted to the interest of their charge, and incapable of any abuse of their large powers. It will be seen how judiciously they used their authority and with what unvarying firmness they upheld the hands of the officers of the Academies, in the maintenance of their scholarship and discipline; not counting the cost, but vindicating the principle of a high standard, whereof the subsequent history of the Institution was to exhibit the solid fruits."[1]

It might appear remarkable to one unacquainted with the history of the times that all five members of the Board appointed by Governor Hammond

---

[1] John P. Thomas, *History of the South Carolina Military Academy*, p. 45.

were generals. The schools, of course, were military schools, designed to replace soldier guards at the two arsenals with cadets, but even so it would seem that some business men as well as military men would be selected to govern their affairs. As a matter of fact, the over-emphasis on the military aspect of the Board was more apparent than real, as they were not professional soldiers, but were all business men of affairs, serving their State as citizens, but also prepared, in those critical days between Nullification and Secession, to serve her as soldiers.

General James Jones was born in Edgefield District in 1805. He graduated from the South Carolina College, studied law, and practiced at Edgefield. In 1836, he served as a captain of volunteers in the Seminole War in Florida, his company forming a part of the regiment of Colonel Abbott H. Brisbane, who was later a professor at the Citadel (1847-1852). After several years' service as Adjutant General of the State, General Jones accepted the position as manager of the first textile manufacturing company in the State, the cotton mill at Vaucluse, South Carolina. During the War, he was for a time colonel of the 14th South Carolina Volunteers, and was later Quartermaster General of the State. He was a recognized authority on militia law. He presided over the destiny of the Citadel until its close in 1865.

General D. F. Jamison was born in Orangeburg District in 1810. He attended the South Carolina College, and studied law, but devoted most of his life to planting. He was considered one of the foremost men of the State. He had a commanding presence, and, being fond of military matters, he rose to the command of a cavalry brigade. He was chairman of the military committee of the House in 1842, and introduced the bill to establish the schools at the Arsenal and the Citadel. When the Secession Convention met in 1860, he was chosen to preside over that historic body. He was a man of literary tastes, and was the author of *Life and Times of Bertrand du Guesclin,* dedicated to his friend William Gilmore Simms. General Jamison contracted yellow fever in Charleston in the summer of 1864, and died in a few hours.

General John H. Means was born in Fairfield District in 1812. He was the thirteenth of fourteen children, one of seven sons noted for their great physical development. He graduated from the South Carolina College in 1832. He was an ardent secessionist, and when the other Southern states seemed reluctant to declare for separation from the Union, he favored State action by South Carolina alone. He was elected governor in December, 1850, when only thirty-eight years of age. Governor Means' term of office was a time of great political excitement. South Carolina spent $350,000 for military arms and equipment, and organized fifteen brigades of infantry and five of cavalry. In April, 1852, there was a State convention to decide if South Carolina should secede. Governor Means presided, and was in favor of secession; but the convention decided that the time was inopportune, and only passed a resolution asserting the State's right to secede. Eight years later, ex-Governor Means was a member of the convention which adopted the Ordinance of Secession, and was foremost among those who took the field to defend the Confederacy. At Second Manassas, he gave his life to that sacred cause. Colonel Thomas says of him: "No custodian of the South Carolina Military Academy in the period before the War exceeded Governor Means in his devotion to her interest. Jones—Jamison—Means—reigned for twenty years, 1842-1862, as the Triumvirate of the School, composing a trinity of guardful forces. To say nothing of the other strong pillars of the Citadel and Arsenal, these three were the Doric columns upon which the State's military edifice rested securely."

General Daniel Wallace was born in Laurens District in 1801. He was a captain of militia at eighteen years of age, and was always fond of military matters, rising to the rank of major general of the 5th Division, South Carolina Militia. He served two terms in Congress. He was sixteen years a member of the Board of Visitors, and it was while in attendance upon a meeting of the Board that he was seized with the fatal illness that terminated his life at his plantation in May, 1859.

General W. J. Hanna was born in 1806 in York District. He studied law and settled in Chesterfield where he established a reputation as an able attorney. He was three times elected senator from his district in the General Assembly, and became a brigadier general of militia. He died at the early age of forty-seven.

General James Willis Cantey, Adjutant General of the State from 1841 to December, 1853, was, by virtue of his office, a member of the Board for eleven years. General Cantey came of a family of military traditions, and was a soldier born. He rose in the militia of the State from the rank of lieutenant to brigadier general. "A gentleman of the old school, and to the manner born, he was the soul of hospitality, and his home was the center of refinement and of every domestic virtue."

Of the three officers composing the first faculty of the Citadel, two had been trained at West Point—Captain W. F. Graham, the superintendent, and Dr. J. E. B. Finley, who had been a cadet at the national military academy for two years, and had later seen service in the Seminole War. Both of these officers died in the spring of 1844, after barely a year's service in the newly organized school.

Captain Graham had developed consumption, and, finding his case without hope of improvement, had tendered his resignation to the Board of

Visitors. Before any action was taken on it, however, death severed his connection with the academy, April 26, 1844.

Colonel J. L. Branch, who was a cadet at the time, writing of Captain Graham's death many years afterwards, says:

"I was the last person who saw him alive. Early in the night, as officer of the guard, I knocked at his room door with a view of giving him the countersign. A deep, hoarse, and almost inaudible voice responded— 'Come in.' Upon opening the door, I approached his bedside, remarking, 'Here is the countersign, Sir.' He laid his arm over toward me, and I placed the little triangularly folded and sealed paper in his right hand, and left the room. It was found next morning unopened. At what hour during the night he died no one ever knew."[2]

At a special meeting of the Board on May 20, Captain Richard W. Colcock was elected superintendent, and Dr. William Hume was elected professor of science to succeed Dr. Finley, who had died on May 11. Dr. Hume, thus elected to the faculty in the Citadel's second year, was to continue as one of its officers until its close at the end of the War of Secession, and this venerable professor was left the sole guardian of the buildings, which he surrendered to the Federal forces on February 18, 1865.

The academic session at the Citadel was unusual in beginning on New Year's Day. From 1843 until 1858 Commencement exercises were held in the latter part of November. During this first period of sixteen years, a semiannual examination was held the third week in April, and was followed by a "military session" of one month during which time no classroom work was done. Similarly, after the Commencement in November, a suspension of studies occurred until the beginning of the new year and the new session. The cadets did not have a vacation, however, for all of these two months, meagre as that would now be considered. The regulation regarding furloughs reads:

"There shall be a suspension of the academic studies for the month immediately succeeding the semi-annual examination, and in December in each year; and the instruction of those months shall be exclusively military. During these months, the Superintendent may, at his discretion, on the applications of parents or guardians, grant furloughs to such of the Cadets as may have deserved such indulgence, for a length of time not exceeding fifteen days. Provided, that such a number of Cadets shall at all times be retained at the Citadel Academy in Charleston and at the Arsenal Academy in Columbia as will be fully sufficient to discharge the duties of

---

[2]In the military escort given the remains of these two Citadel officers, the corps of cadets was assisted by the Washington Light Infantry. Thus early had a comradeship been established between these military organizations which ripened as the years increased and wielded an influence on a number of important occasions.

public guard for the arsenals efficiently,[3] and no leave of absence shall be given at any other time except in cases of strong necessity."

The restricted accommodations of the buildings in the early years of the academies necessarily limited the size of the student body. The capacity of the Citadel barracks for the first ten years was seventy-five cadets, and of the Arsenal, thirty-three.

In 1842, the Citadel consisted only of the quadrangular building in the center of the present group, and of only two stories. There was no gallery on the north side of the quadrangle, but an interior narrow passage known as The Lane, with cadet rooms on each side. In 1849, the third story was added to this building, and The Lane abolished, the gallery on the north side being constructed with arches to conform to the other sides of the interior. It was not until 1854 that the two wings were built.

Fifty-four cadets were to take the place of the soldiers who served as guards at the Citadel and the Arsenal; and as they were to perform all the duties of these guards, they were to be "maintained and educated at the public expense." These fifty-four beneficiary, or State cadets, were apportioned among the twenty-nine districts of the State in proportion to their population, and to be "selected from those not able to bear their own expenses."[4] In addition to these, it was proposed to admit the same number of pay cadets apportioned in the same way among the several districts.

The fees charged the pay cadets were fixed at $200 a year, and remained at this remarkably low figure until the high prices of the war times made it necessary for the Board to raise them.

It may be assumed that at first this sum covered the entire cost of a cadet's maintenance, including "subsistence, clothing, tuition, books and stationery, medical attendance, and all charges and expenses whatever for the year."

But in 1854, the Board estimated that the actual cost per cadet had increased about seventeen per cent. In their report to the General Assembly for that year they state: "When these Institutions first went into operation, the annual amount for the support of a pay cadet was fixed at two hundred dollars, and that amount was deemed sufficient until within a recent period; but the high prices of provisions and labor during the past year have raised the sum required for the maintenance of a cadet to two hundred and thirty-eight dollars. Upon that condition of things, it was conceived by the Board whether it was not advisable to increase the amount required of a pay cadet to that sum; but they have concluded not to make the increase, at least for the present, in the expectation that prices may decline, but more

---
[3] See Inventory, p. 171, Minutes of the Board of Visitors.
[4] Minutes of the Board of Visitors, p. 103.

especially as they are unwilling to exchange that distinctive feature of these institutions which has made them cheap schools for that large portion of the people of the State who are unable to afford their children a more expensive education; and because the cadets make a return in part for their education and maintenance by acting as a guard to the public property at the State arsenals. The Legislature may think proper to make some addition to the annual appropriation for the support of these Academies to enable the Board to accomplish the end they have in view without an increase of the annual amount required of the pay cadets."

The knowledge requisite for admission to the lowest class, called the Fourth Class, was no higher in the *Regulations* than ability "to read and write with facility." As a matter of fact, however, the faculty conducted an examination of the applicants for admission, and rejected those whom they considered incompetent to undertake the course of studies. Statistics show that up to the year 1855, fourteen per cent of the candidates appointed to the academies "failed to report or were rejected." Candidates were required to be between the ages of fifteen and nineteen years, and to furnish certificates of recommendation from the schools which they had attended, so that the standard was probably not much below that required in other institutions of higher education in the State.

While the act establishing the schools at the Citadel and the Arsenal had in view, without a doubt, two separate and distinct institutions, the fortunate fact that one board of visitors was appointed to administer the affairs of both academies led at once to the adoption of a plan of coöperation whereby the instruction given at the Arsenal was limited to the first year's work, after which the cadets went to the Citadel for the other three years' courses of study, thereby avoiding a useless duplication of scholastic work, and effecting an economy of administration.

The textbooks used for the recruits at the Arsenal during the first year were Arithmetic (Davies), Algebra (Davies-Bourdon), English Grammar (Murray), History of South Carolina (Simms), and French Grammar and Reader (Collott). In 1856, some changes had been made in the texts used, Loomis' Algebra and Adams' Arithmetic having supplanted Davies, Bullion's English Grammar having replaced Murray's, and Willson's History of the United States having been substituted for Simms' History of South Carolina. Some additional studies were also taken up—Morse's Geography, Mythology, and Landscape Drawing appearing among the studies of these Fourth Classmen. From 1861 to 1865, Bookkeeping was also added.

The curriculum at the Citadel, beginning with the second year, consisted of Geometry and Trigonometry (Davies), Descriptive Geometry,

Surveying, and Analytic Geometry (Davies), Universal History (Tytler), French (Histoire de Charles XII), and Tactics (Scott). In 1860, a course in Elocution and Composition was added to this class. This Third Class, as it was called, averaged about thirty-three members.

In the Second Class (Junior), Church's Calculus was the text in Mathematics taught during the whole period from 1845 to 1865. Courtenay Boucharlat's Traité de Mécanique, Bird's Elements, and Gummere's Astronomy were texts in applied Mathematics. Boucharlat gave place to Bartlett's Mechanics in 1851, and later Muller's Principles of Physics took its place. Fowne's Chemistry and Dana's Mineralogy were the science texts studied by this class. Blair's Rhetoric held its position in the curriculum of this class throughout the whole period, Shaw's History of English Literature being added to the course in Belles Lettres in 1856, in which year, also, the study of Logic was introduced, with a textbook in French. Linear drawing, with Shades, Shadows, Isometric and Perspective, and both infantry and artillery Tactics also formed part of the course. This class numbered about twenty-five members on the average.

Civil and Military Engineering, with texts by Mahan and Halleck, were taught in the First Class, or senior year. In Belles Lettres, Abercrombie's Intellectual and Moral Philosophy, Wayland's Political Economy, and Story's National and Constitutional Law were the texts used in 1849. Upham's Intellectual Philosophy and Paley's Moral Philosophy took the place of Abercrombie's text in 1851, the latter being replaced in 1861 by the same author's Evidences of Christianity, and in 1863 by Butler's Analogy. A book in Political Economy in French, by Droz, was used instead of Wayland's Elements after 1855. The constitutions of the United States, and of South Carolina, and Calhoun's Disquisitions on Government were studies of special importance during the decade preceding the War Between the States. Vattel's International Law was also added in 1860. In the department of experimental sciences, geology was added to advanced courses in chemistry and mineralogy. Architectural drawing, and both infantry and artillery tactics were also studies pursued by this class, which varied in size from seven to twenty-six members.

A consideration of this curriculum, all the subjects of which had to be passed by all the members of the cadet corps—for there were no electives —will no doubt justify in the reader's mind the quotation from Milton which Colonel Thomas, in his *History of the South Carolina Military Academy*, appropriated as a fitting characterization of the course of studies pursued by the cadets: "A complete and generous education: that which fits a man to perform justly, skillfully, and magnanimously all the offices of a citizen, both private and public, of peace and war."

That the course of studies was not easy for a boy of inferior intelligence, or not of studious habits, was indicated by the large percentage of failures. In the first ten years statistics show that of the beneficiary, or State cadets who matriculated, only thirty-one per cent completed the course successfully and graduated. Of the pay cadets, fewer still—eighteen per cent—graduated. The Board of Visitors, in its report to the General Assembly, December 3, 1853, comments thus on the excessive number of casualties:

"This exhibit shows that of the number appointed to the institution a large portion are discharged or dismissed before the termination of their enlistment. That is owing in a great measure to want of care on the part of parents, guardians, and commissioners of free schools in selecting young men for the Academies, who, from previous imperfect instruction, improper moral training, or natural incapacity are either rejected on their presentment for admission, discharged after the end of their probationary term of four months, or they break down at some subsequent period of their four years course. Besides, they are subjected to a rigid system of instruction, and a severe course of study which requires some talent and much diligence to withstand and complete, and the boy who can not work, or who will not work, is discharged, and his place is given to another.

"But the difficulty of completing the course of instruction, and the large number of dismissals in this institution, so far from being an argument against the system pursued, they are, in the opinion of the Board, not the least cause of its excellence. It is impossible to educate everybody; and it seems to be better, by fixing a high standard of education, to train up those who are subjected to it, so as to enable them to accomplish the greater amount of good, than to educate a large number imperfectly; for, if the standard is low, very many will fall short of even that mark; and the tendency of such a system will be to deteriorate until it becomes worthless, and in the end contemptible."

Again, in 1854, commenting on the fact that the graduating class numbered only thirteen out of sixty-eight who had matriculated in 1851, the Board says:

"It will be perceived that the number of graduates is small when compared to the number of which the class first consisted; but of the number admitted, all received instruction in a greater or less degree. A portion of the class, from want of capacity or industry, was honorably discharged or dismissed at the end of the probationary term of four months.

"Some of them held on to the end of the first year; others as they advanced to the higher branches, finding difficulties which they could not surmount, successively fell off, until upon the completion of the course, but thirteen (of the original sixty-eight) were found worthy of the diplomas conferred under the regulations of the institution.

"It may appear at first view that the standard has been fixed too high; but the Board believe all who have not been dismissed for misconduct have been educated to the extent of their capacity, and they indulge the hope that, though not found entitled to the highest distinctions of the institution, very many of them have been thus prepared for useful and honorable stations in life."

Cadet Life

A cadet at the Citadel led a busy life. His conduct was governed by military regulations, some in minute detail, and he had little time for idleness or dissipation. His day's work began at six o'clock, which, on winter days, was about dawn, and ended at half past nine at night during the winter months, half past ten when the days were longer.

The regulations for the year 1849 give the following general roll calls, which have remained almost unchanged down to the present:

Reveille—Will be regulated by the beating of the Picquet Guard.

"Peas Upon a Trencher"—Signal for breakfast—8 A.M., March to September; 7:30 the rest of the year.

"Roast Beef"—Signal for dinner, 1 P.M.

Retreat—At sunset.

Tattoo—10 P.M., March to September; 9 P.M. the rest of the year.

The Surgeon's Call was sounded daily at 6:30 A.M. for those needing medical attention. Taps, for lights out, was sounded half an hour after tattoo.

Study hours in barracks were observed from reveille to breakfast, and in the evenings from half an hour after supper until tattoo. The recitation periods were from 9 A.M. to 1 P.M., and from 2 to 4 P.M., except on Saturdays and Sundays, on which days no recitations were held.

The *Regulations* prescribed that "every cadet, on rising in the morning, shall roll up his mattress, with the bed-clothes neatly folded in it, put it into the bed-sack and strap it. He shall hang up his extra clothing, put such articles in the clothes-bag as it is intended to contain, clean his candlesticks, and arrange all his effects in the prescribed order."

The cadet rooms were large and could easily accommodate four occupants. The inmates took turns by the week in performing duty as room orderly. It was the duty of this cadet to sweep out the room fifteen minutes after reveille, attend to the making of fires when necessary, see that lights were in the room ten minutes after call to quarters and that they were extinguished at taps. Every other Saturday, the cadets in every room washed the floors, washstands, mantels, and shelves, and prepared them generally for inspection.

From March 1 until December 1, there was an infantry or artillery drill held each day except Saturdays and Sundays. On Saturdays, besides the inspection of rooms in barracks, there was an inspection under arms of the company of cadets, and on Sundays all cadets were required to attend church.

An examination of this schedule will show that there was not a great deal of time for a cadet to enjoy his freedom in "release from quarters"— about half an hour after breakfast, dinner, and supper, and a period of variable length depending on the season of the year in the afternoon before retreat. At these times the cadet had the liberty of the Green. In the early years, there were no organized athletics, and the following paragraph in the *Regulations* sounds quaint to modern ears: "The recreation ground for athletic exercises, such as walking, jumping, running, and playing ball (in all of which the Superintendent would recommend practical exercise), will be bounded by the fence surrounding the Green."

Visiting in the city was governed by regulations which would now be considered remarkably strict: "No cadet shall go into Town at any time, except with a written permission of the Superintendent, or of the commanding officer for the time being, and such permit must specify the object of the visit and the name of the person or place to be visited."

"No cadet shall visit a family or families in Citadel or Town at any time, except it be by written permission of the Superintendent, and upon the special invitation of the head of the family."

And, most inclusive of all: "No cadet shall receive or apply for permission to dine or take any meal, or sleep at any hotel, or any other place than the Citadel, as prescribed in the regulations."

In the absence of intercollegiate contests, so characteristic of present day college life, the cadets of the Citadel in the fifties found a vent for their inherent spirit of rivalry and youthful enthusiasm practically only in the debates and oratorical contests between the two literary societies in the corps. From all accounts of the ante bellum alumni, the rivalry between the members of the Calliopean and Polytechnic societies was astonishingly bitter; so acrimonious, indeed, that the manly art of self-defense was promoted concomitantly in order to settle issues that arose without relevancy to parliamentary law. Minerva and Mars took equal delight in the contests of these young soldiers.[5] A third society, the Amosophic, was born but never reached maturity! (Minutes of the Board of Visitors, p. 117.)

---

[5] In a letter from Charleston, written February 2, 1855, a Northern woman travelling in the South, says:

"I visited the Arsenal and Citadel Academy, and was invited by Major Capers, superintendent and professor, to walk through the establishment. It is modeled after the plan of West Point, and equals that institution in all respects except in languages. Here are taught but two foreign languages, French and Spanish. The chambers of the two debating societies are richly fitted, containing beautiful banners bearing appro-

Gen. C. I. Walker, Class of 1861, says:

"In our day there were no athletic interests in the Citadel and but few in any of the colleges of the country. Our great interests were centered in the two Literary Societies, between which there was great rivalry. The difference between the membership of the societies was largely social and geographical. The Calliopean got the men from the low country and the Polytechnic those from the up country. At times the feeling ran so high that a man of one Society would not enter the room of men from the other, without stopping to knock. The Calliopean Hall was in the room at the northwest corner of the west wing, and that of the Polytechnic just east of the same. Both were fitted up nicely from the private funds of the Society. The President was elected quarterly and taken from the first class. The special honor was to be elected President for the first quarter. I had that honor in my class."

Colonel Coward, writing in 1910 about "Life at the Citadel before the War," says:

"We had no social functions such as hops, picnics, and class banquets given by cadets. We had, however, a Fourth of July Association, which held an annual celebration at which the Declaration of Independence was read, and an address made by a member of the First Class.

"A platform was erected in front of the flag-staff and covered with a canopy of tent-flies for the speakers and invited guests, and all the class-room chairs and benches were arranged under the galleries for seating the general public. In our estimation, this was a grand occasion that warmly aroused our spirit and pride."

The Colonel was looking backward nearly sixty years, and the retrospect made him philosophize, for he quotes, aptly enough:

"Tempora mutantur, et nos mutamur in illis."

In the middle of the nineteenth century, a college commencement was taken very seriously.

When the time came around for the first graduating exercises of the Citadel, November 20, 1846, the custom of the other institutions of higher education was followed, and elaborate preparations made for the important event.

The first feature of the celebration was the procession. This was—on *paper,* at least—a parade of dignitaries and societies well worth seeing.

The advertisement stated that the procession would move from the

---

priate inscriptions, a handsome lecture room, etc., etc. On entering the armory (where secession is locked in, but possessing a ready key), the major pointed out two brass field pieces bearing Turkish arms, taken from the Turks by Commodore Decatur, and sold by Mrs. Decatur to the Arsenal of Charleston." *Wayside Glimpses, North and South,* by Lillian Foster.

Citadel at 10 o'clock, A.M., and proceed to the Second Baptist Church[6] on Wentworth Street, where the exercises would be held.

The order of the procession was as follows:

Marshals
Major W. Allston Pringle
Lieutenant H. C. King
Cadets Jones and Hagood
The Cadets
Students of the Medical College
Citizens generally
Strangers, visitors to the city
The Mercantile Library Society
The Apprentice's Library Society
The Charleston Library Society
The '76 and Cincinnati Society
Commissioners, Stewards, and Instructors of the Orphan House
Commissioners of the Free Schools
The South Carolina Society
Board of Supervisors, Principals, and Teachers of the High School
The Board of Trustees and Faculty of the College of Charleston
The Board of Trustees and Faculty of the Medical College
Field and other Officers of the 4th Brigade
Brigadier General and Staff of the 4th Brigade
Major General and Staff of the 2nd Division, S.C.T.
Officers U. S. Army and Navy
Members of Congress
Foreign Consuls
United States and State Judges
Clergy of all denominations
Officers and Professors of the Citadel Academy
The Governor's Staff
His Excellency, the Governor, and other members of the Board of Visitors

The exercises at the church consisted of a prayer by Reverend Mr. Kelly and addresses by members of the graduating class. Cadet John H. Swift was excused from delivering his oration on "Genius" on account of having given on the evening before a lecture at the Citadel before the Board of Visitors and other prominent citizens on "The Morse Telegraph," then a novelty. All the other members made addresses, and two of them made two—Charles Courtenay Tew, the first-honor graduate (and therefore the

---

[6] Now the house of worship of Centenary Church, colored.

proto-graduate of the Citadel), speaking on "Prejudice and Passion," and later delivering the valedictory address to his class; Richard G. White, second-honor man, making the salutatory address and another on "Industrious Efforts." The other orations were by John L. Branch, "Materialism Inconsistent with Philosophy"; Charles O. Lamotte, "Public Opinion," and William J. Magill, "Conscience and Innate Moral Sense." These addresses were interspersed with music, and the exercises closed with the presentation of diplomas to the graduates by the chairman of the Board of Visitors, and some closing remarks by the superintendent, Major Colcock.

The custom of having only cadet speakers continued for a number of years, but when the number of graduates increased, all, fortunately, were not required to make orations. In 1851, for instance, when a class of twenty-six was graduated, the endurance of the audience would obviously have been overtaxed if this practice had been continued. As it was, only four cadet orations were delivered at this commencement other than the salutatory and valedictory addresses.

In 1854, the address of Cadet Asbury Coward, on "Woman's Rights," was selected by the press for special commendation.

The next year, a departure was made from the custom in vogue, and the feature of the occasion was an address by Colonel I. D. Wilson, followed by the Alumni Association speaker, F. F. Warley, Esquire, of Darlington. On the evening of commencement it was customary for the Polytechnic and Calliopean Societies to hold their commencement exercises and to have some distinguished speaker to address them. (For a list of these speakers see Appendix G.)

There were no graduating exercises in November, 1853, due to an émeute of the Second Class in the preceding year, as a result of which all the members of that class left the Citadel in a body. This will be referred to again later. The two literary societies, however, celebrated as usual their anniversary, on Wednesday evening, November 23, with an oration by Honorable Richard Yeadon, who took as the subject of his address: "The Necessity of Subordination in Our Academies and Colleges, Civil and Military," a topic the timeliness of which he called to the attention of the cadets in these words:

"My ambition on this occasion is not to win the laurel of eloquence in the service of your Muse, Calliope, and her votaries; but my humbler and, I trust, more useful aim shall be to impress your minds and hearts in the way of plain and friendly admonition on a subject of the highest importance to the youth of our State in general, and to you in particular—one, too, to which recent events in the South Carolina College and in the Military Academy, of which you are pupils and members, have given a more than ordinary prominence and interest."

In the account of the exercises of the Commencement of 1855, the *Courier* on the day following contained the following paragraph:

"The Citadel Academy is second to the State College but in age and in number of alumni. In no other points can it be considered inferior or less worthy of attention, and still less can it be regarded as a rival. Some, we know, would reduce all public instructions to diagrams, and the juvenile democracy of letting every boy study when, where, and how he pleases, have branded it as anomalous, and others have made sundry suggestions of change, to avoid what is regarded the wasteful expenditure of educating thoroughly a comparatively limited number. It is, however, becoming less important every day that such carpings and cavillings should be answered. The Citadel Academy is steadily and effectively sending forth its 'living epistles' of argument, appeal, and vindication, and these are being read of all."

The first graduating exercises of the Citadel in 1846 were held in the midst of the excitement incident to the beginning of the Mexican War. South Carolina was preparing to furnish her quota of the volunteer army which the president was going to despatch to Mexico. This was the Palmetto Regiment—an organization that performed its full duty in the campaign south of the Rio Grande. Already, the Citadel had made some reputation as a "school of arms," and the recruits for the regiment were sent to Charleston for military instruction under the cadets as drillmasters.[7]

---

[7] The venerable Dr. Divver, of Anderson, South Carolina, writing June 4, 1925, at the age of eighty-five says: "In the early forties, my father resided in the city of Charleston in a brick house on King St. The house is still standing directly opposite the middle of the Citadel Green. Being at that time quite a small child, my negro nurse would take me out on the Green to see the new cadets drill. This was the first class at the organization of the Citadel, the fall of 1843. I remember C. C. Tew, C. O. La Motte, and in the next class Sam Jones, Johnson Hagood, and others.

"I also saw a lieutenant from Fort Moultrie, who, with other visitors, attracted my attention, as he visited my father's home opposite the Citadel to talk to my mother about her relatives who moved to Ohio in 1810. His name was W. T. Sherman. The next time I saw General Sherman was on a railroad train going to Atlanta in 1881 by invitation of the big Exposition authorities of that city. Strange to say, I recognized the man.

"I also witnessed the Palmetto Regiment tented on the Citadel Green, and their review by Governor David Johnson just before they went to Mexico."

The following extract is taken from an article in the *Southern Quarterly Review* for November, 1850:

"Who does not remember the Charleston company of the famed Palmetto Regiment when camped on the Citadel Square? Who does not know the modest offer of the cadets, through their lieutenant commanding, to initiate the company in the rudiments of infantry tactics; and who has forgotten the stately though juvenile forms of the Cadets marching and commanding their seniors with all the confidence of veterans?

"The officers very properly embraced the offer; and while the men were drilled on the Square, they themselves underwent like drilling on the more private muster ground of the Citadel.

"On the arrival of other companies and their encampment at the Magnolia Farm, the Cadets still continued the daily drills, so that each company had the advantage of instruction from a practical teacher.

In the Palmetto Regiment was a boy of sixteen years named Allen H. Little, who went to Mexico and lost an arm in one of the battles of the War. Upon his return to South Carolina, he was awarded a scholarship to the State military academies, but on account of inadequate previous schooling, it took him two years to complete the Fourth Class at the Arsenal. When, however, he was promoted to the Citadel, he showed himself to be a persevering and ambitious student, and stood at the head of his class during the last two years of his course, graduating with first honors in 1852.

It was his intention to study for the bar, "but," in the words of Colonel Thomas, "no ordinary physical frame could stand the strain to which his mind subjected its tenement. He sickened, and in less than a year after entering upon life's career, he died in the bloom of manhood.

"Among that band of Citadel graduates, on whom, at their humble birth, fortune smiled no propitious radiance, the name of Allen H. Little is conspicuous. He is presented in these pages as one of the fruits of the crowning feature of the Academy—a system which, under an eclectic device, takes the poor but deserving boy of the State and opens to him the avenue of higher education."[8]

Allen H. Little was not the Citadel's only veteran of the Mexican War. Another beneficiary cadet was admitted to the Citadel about the same time that Little entered, in this case, also, on account of services rendered as a soldier in the Mexican War. This boy had been severely wounded in the leg, so that he was exempted from military duty as a cadet. The career of this boy was, however, in marked contrast to that of Allen Little, for, although mentally bright, he would not study, and was discharged for deficiency. He lived long, and had a picturesque and checkered career. In "radical days" in South Carolina after the War, he was known as Judge Thomas J. Mackey.

In 1881, when the Board of Visitors were seeking to obtain possession of the Citadel from the United States government, they engaged Judge Mackey as a special agent in Washington to push the claim, and it must be said to the credit of the judge that he was zealous in the service of his Alma Mater, although his efforts were not crowned with success.

In April, 1850, the Citadel Green was the scene of a solemn and memorable ceremony.

Senator John C. Calhoun had died in Washington on March 31 of that

---

"How far beneficial this drilling ultimately was we cannot say; but we have pleasure in claiming for the Citadel Academy the honour of having first drilled the Palmetto Regiment; and we claim that a portion of their after fame should redound to the military school which gave them the first lessons in the art of combining individual gallantry into a phalanx of irresistible power."

[8] Colonel J. P. Thomas, *History of the South Carolina Military Academy*, p. 84.

year. His body had been kept in the capital for several weeks, and was brought to Charleston the latter part of April, being carried by rail to Wilmington, North Carolina, and then transported under escort by the ship *Nina* to Charleston, a special platform having been erected on the quarterdeck of the vessel for the casket. At Charleston, impressive ceremonies were arranged for receiving the body of the great Carolinian. At the foot of Boundary (now Calhoun) Street, the casket was transferred from the *Nina* to a specially made car modelled after the funeral car of Napoleon, and drawn by six horses caparisoned in mourning, with grooms in deep black. Six ex-governors and lieutenant-governors acted as pallbearers, and there was a military escort consisting of the Washington Light Infantry, German Fusiliers, and Marion Artillery. The cortege passed through Charlotte Street to the Citadel Square, where Governor Seabrook waited to receive the Congressional delegation in charge of the body.

The Citadel and its parapet were appropriately dressed with mourning streamers and rosettes, and other funeral decorations, forming an appropriate background for the exercises that took place.

Senator Mason delivered the body to the governor in an appropriate address, and the governor replied fittingly, committing the remains in turn to Mayor Hutchinson. A procession was then formed, passing from the Citadel Square down King Street to Hasell, thence down Meeting Street to the Battery, around through East Bay and Broad Streets to the city hall where a catafalque had been specially prepared to receive the casket. Here it lay in state for two days before its interment in St. Philip's churchyard.

The chief ornament of the Square at the present time is the imposing monument erected in 1887 to South Carolina's greatest statesman.

Yellow Fever

The scourge of yellow fever, so prevalent in tropical countries during the nineteenth century and before, visited Charleston at intervals, appearing always in the summer months or early fall; and as there was no vacation at the Citadel at this season, the menace to the cadet corps was of serious concern to the authorities.

In the summer of 1849, Major Colcock, superintendent of the Citadel, disbanded the corps on account of the prevalence of the fever in the city. It appeared that he acted without consulting the Board of Visitors, for, at the meeting of the Board on November 30, of that year, a resolution of disapproval of this action was passed; and certain cadets who did not run away from the fever were commended, and were granted a furlough of one month.

In the summer of 1852, a considerable part of the cadet corps was seized with panic, or tempted, by the opportunity offered by the appearance of yellow fever in the city, to take an unauthorized vacation from the Academy. They were taken back later, but not without a caustic reprimand from the Board, which passed the following resolution at its meeting on November 30:

"The Board of Visitors, in permitting such cadets of the Citadel Academy who deserted their posts upon the appearance of the Yellow Fever during the past summer to return to their duties, do so in the hope that they will not be called upon again to animadvert upon conduct which strikes at the root of all discipline, and gives ground for apprehension that they who prefer safety to duty are not the best Guardians of the public property.

"The Board are willing to believe that the disreputable panic which led to the desertion of the first of September was not so much the result of a wilful abandonment of duty on the part of the Cadets as an improper interference and unwise counsels on the part of others; and they trust that upon the occurrence of a similar event, the Cadets of the Military Academy will not place their security in flight, but in the hands of those responsible for their safety and the well-being of the institution of which they form a part."

On this occasion, again there were some who stuck by the colors, and were commended by name in orders, which read:

"*Resolved,* that all demerit marks up to the time of the Yellow Fever be struck out against the following cadets, viz., Brice, Dubose, Lee, Munnerlyn, Prior, Snowden, Venning, Wylie, R. Young, in consideration of the high estimate the Board put upon their conduct for remaining at their post until furlough."

Two years later, in 1854, Yellow Jack visited Charleston again, and an unprecedented number of persons were attacked. It was estimated that upwards of 20,000 cases of the disease occurred. The editor of the *Charleston Medical Journal and Review* spoke of it as "not an Epidemic, but a Pandemic." The fever, however, was not of a very virulent type, the number of fatal cases being 621, or about three per cent.

In the annual report to the General Assembly, the Board had this to say about the epidemic:

"The Board regret the necessity which was imposed upon them of disbanding a great portion of the Corps at the Citadel on the occurrence of Yellow Fever during the past summer. When the disease was first known to exist in the City of Charleston, the Chairman, acting under the direction and with the approbation of the Board, obtained from time to time reliable reports from the most eminent physicians of the City, and at length upon their advice and with the concurrence of the surgeon of the post, after leaving a sufficient number of acclimated Cadets to serve as a Guard, he ordered the First Class with their professors to repair to the Arsenal at Columbia, and disbanded the remainder of the Corps until fever should cease as an epidemic."

In the same report, the Board refers to some complaint from the merchants of Charleston on the action which they had taken, but stated that while the Board "would commit no act which would unnecessarily prejudice the commercial interests of the City, they would not incur the responsibility of the loss of a single life entrusted to them which care and precaution could prevent."

In the investigations and discussions as to the origin of yellow fever, Dr. William Hume, professor of experimental sciences at the Citadel, took a prominent and important part.

In the spring of 1854, before the advent of the epidemic of that year, he had submitted to the city council of Charleston a report on the "Source and Origin of Yellow Fever and Such Means as May Seem Worthy of Adoption in Order to Obviate Its Future Occurrence." In this report, he gives an account of the thirty-four visitations of this scourge at irregular intervals from 1699 to 1854, and argues a connection between Charleston's intercourse with the West Indies and the epidemic years, arriving at the conclusion that yellow fever was an importation from the tropics, and that it could be prevented by a proper quarantine. He believed so firmly in this thesis that he becomes rhetorical. He says: "Shall we receive and worship it (the fever) as a monarch, or shall we expel it as a tyrant?"

Again, in January, 1855, after the fatal year of 1854, he contributes an article to the *Charleston Medical Journal and Review* in which he specifically states: "We trace it (the fever of 1854) three separate times to West India vessels, and at three separate periods, viz., on the 11th of May, on the 11th of July, and on the 26th of July."

The scientific men of the day were not at all agreed as to the cause of the disease, and their theories were almost as widely discordant as the popular superstitions on the subject. Efforts were made to trace a connection between the epidemics and every known meteorological condition about which data had been collected. An endemic origin was generally believed in, assigning as a cause the filling in of marsh lands with organic garbage, or the opening of street drains in the summer time; while some attributed it to "undiscovered pestilential gases," "unknown germs in the soil," or "moisture in the earth"! Speculations were made why epidemics followed years in which conflagrations had occurred in the city; and some firmly believed that the habitat of the fever was the Spanish moss which grew so luxuriantly in the coastal country. A hundred guesses were made as to who was the culprit responsible for the scourge, but not one suspicion seems to have existed about the moral character of *Aedes Aegypti,* the common little mosquito that buzzed and bit the inhabitants during the summer season. It took another half century more of research by the scientific detectives to expose, convict, and put this deadly pest behind the (mosquito) bars.

Again in the summer of 1856 yellow fever appeared in Charleston, and the Board once more deemed it advisable to transfer the Citadel cadets to the Arsenal, although the quarters there were wholly inadequate for the accommodations of both corps.

Since the establishment of the schools, yellow fever had menaced the safety of the Citadel in five different years—1843, 1849, 1852, 1854—and now in 1856.[9] The Board began to wonder if these enforced migrations must be expected indefinitely; and if so, what policy should be adopted in the circumstances.

The next year, at the annual meeting on December 1, 1857, they made an important change in the academic year which was induced largely by the fever problem. In their report to the legislature they say:

"The Board have heretofore felt the necessity of extending the academic term beyond the period of four years, as that time has been found too short to complete certain branches taught during the last year of the course, and as the almost yearly recurrence of yellow fever in the city of Charleston has forced upon them a change in the periods of the military sessions from April and December to the months of August and September, they have also changed the time of the annual examinations and Commencement at the Citadel from November to April, thus extending the whole period of instruction to four years and three months."

Due to this change, the class which would have normally graduated in November, 1858, received their diplomas in April, 1859, and no class of 1858, consequently, appears on the roll of graduates of the Academy. It will be noted that no class of 1853 appears on the roll; but the hiatus in this case was for a very different reason.

It was a disastrous case of discipline which occurred in the summer of 1852, involving the entire Second Class—the class which was due to graduate in November, 1853.

A personal difficulty had occurred between the adjutant, who was a member of the First Class, and the leading man of the Second Class, as a result of which the latter was suspended. The entire Second Class,

---

[9] In the summer of 1864 General D. F. Jamison, "one of the strongest and most polished pillars of the institution," fell a prey to the grim disease.

thirty-four in number, took up the cause of their suspended classmate, believing that an injustice had been done him, met, and passed a resolution that they would leave the Academy in a body if the order of suspension was not revoked.

This action was clearly in violation of Paragraph 82 of the *Regulations*, which read as follows:

"82. All combinations, under any pretext whatever, are strictly prohibited. Any Cadet, who, in concert with others, shall adopt any measure under pretense of procuring a redress of grievances, or enter into any agreement with a view to violate or evade any regulation of the Academy, or shall endeavor to persuade others to enter into such concert or agreement, shall be dismissed."

The ultimatum of the Second Class was met with a refusal from the superintendent, Major Colcock, who warned the cadets that their action was punishable with dismissal, whereupon "the aggrieved class went out of the Academy, carrying in their arms the little sentinel and his gun, set him down on the Green, called the roll of the class, and, all being present, they dispersed into the city without disorder of any kind."[10]

"It was the old story of hasty action on the part of spirited youth." Such is the comment of the historian of the Academy, Colonel Thomas.

The *Regulations* provided for a redress if any cadet should consider himself wronged by another, or by a professor or officer, by an appeal to the superintendent, or, beyond him, to the Board of Visitors.

The impatience of youth, however, prefers ways of its own to the orderly processes of law.

The minutes of the Board of Visitors for the annual meeting held in the governor's office on November 26, 1852, contain no reference to this unfortunate episode, recording only the dismissal of thirty-seven cadets specified by name. But a year later, December 3, 1853, in their report to the General Assembly, they say:

"The Board regret to state that no class was presented to them for graduation this year. That circumstance was owing to an unhappy outbreak of the Second Class which took place during the summer of 1852 which resulted in the suspension of the whole class; and as the question of their restoration involved the future discipline and government of the Military Academies, the Board of Visitors did not hesitate to confirm the suspension imposed by the Superintendent."

---

[10] Statement of Colonel Coward, who was a Third Class man in the Academy at the time.

CHAPTER IV

FOURTH DECADE: 1852-1862

THE year 1852 was not a happy one for the Citadel. In the summer, there had been the panic among the cadets caused by the yellow fever, to which reference has been made; then followed the émeute in the Second Class, resulting in their expulsion; and finally at the meeting of the Board of Visitors in November, all five members of the Academic faculty resigned for reasons not recorded. The resignations of Major R. W. Colcock, the superintendent, and Lieutenant J. A. Leland, professor of mathematics and astronomy, were accepted; those of Dr. William Hume, professor of experimental science, and F. Gauthier, professor of French and drawing, were not. In the case of Captain A. H. Brisbane, professor of belles lettres and ethics, it appears that the Board exercised some academic discipline and removed him without the formality of a resignation.

Colonel Thomas, whose four years' cadetship was under Major Colcock, gives this characterization of him: "Major Colcock was in the service of the Citadel Academy for eight years. His bearing was that of a soldier. Beneath a somewhat stern exterior he carried a kind and sympathetic nature. Cadets who came nearest to him in the duties of the Academy appreciated him most. He was no speaker, and it was not his habit to indulge in appeals to the corps. His disciplinary methods consisted in the steady application of the law, carried out in military fashion. Bringing to the discharge of the duties of his office a soldier's training derived from his cadetship at West Point and from his experience in the United States Army, he gave to the Academy a military tone and impress that contributed largely to the development of its fortunes, especially as a school of arms."

Captain Brisbane was also a West Point graduate, in the class of 1825, a year before Major Colcock. He served only a few years in the Army, and had a picturesque and varied career. Colonel Thomas describes him as "a remarkable man. A more earnest worker it would be hard to find. His ardent temperament, his volcanic outbursts of convictions, and his fiery zeal in letters will long be remembered by those students of the Citadel who passed under his instruction. There was a martyr-like element in his composition, and for the truth as he saw it he would willingly have imperilled fortune or even laid down his life. In character, he was generous and chivalrous; as a husband, tender and true; as a man, straightforward, brave, outspoken."

That the discipline of the military schools, patterned after the regulations of the national academy at West Point, was consistently administered is evident from an examination of the statistics. In the first decade of their existence, there was not a year in which there were not from four to fifteen dismissals for serious violations of regulations. Of the 550 candidates who were admitted during this period, twenty-two per cent failed in studies, and twenty per cent were dismissed for misconduct.

Early in the history of the schools, the authorities had an occasion to announce clearly their position on the matter of discipline. In a case of insubordination which occurred at the Arsenal in September, 1845, in which the superintendent, Captain Herbert, suspended a cadet for "positive disobedience of orders and mutinous conduct," the Board confirmed the suspension by dismissing the cadet, and stating in orders that the action of the superintendent "was not only proper, but a most necessary and indispensable act of duty."

The discipline at the Arsenal must have been generally disappointing, for the Board, in its report to the governor, December 2, 1845, after commending the work of the Citadel, goes on to say:

"The Arsenal in Columbia has not succeeded in accomplishing the wishes of the Board of Visitors to the extent desired. *The cause of this failure is mainly, if not entirely, referable to the want of proper physical means to enforce the necessary rigid discipline and accountability of the Cadets.*

"If the maintenance of this institution, either as a school or as a depot of arms, was demanded by necessity or the convenience of the country, the Board would ask your Excellency to urge upon the Legislature the necessity of granting an appropriation sufficient to erect a strong wall enclosing the post, and to pay additional instructors. But believing that the increased means and diminished expense of transportation has satisfied all the reasons for the establishment of the post as an arsenal, and that the Citadel in Charleston possesses advantages for a school which can not be realized here, we respectfully recommend that the two institutions, both as arsenals and schools, be consolidated at the Citadel."[1]

---

[1] One sentence in the report above is italicized in order to call attention to an opinion which has prevailed from 1845 to the present time. The Citadel was fortunate in having for its home a building constructed for a fortification and peculiarly well adapted for preserving discipline. When the sally-port gates were closed at night, the superintendent had a feeling that his little band of boys were safe and sound—all practically under the eye of the cadet guard—and there was little anxiety about them being abroad in the doubtful haunts of the city.

Eighty years afterwards, when in 1920, architects planned the barracks for the greater Citadel, the alumni urged with practical unanimity that the quadrangular form of building which had been so effective at the Old Citadel, be preserved at the new plant. To the surprise of some educational critics, who knew nothing of the "genius of

While the recommendation of the Board that the Arsenal be abolished was not acted upon, it was entirely within the Board's power to make it subsidiary to the Citadel, and this was done, the function of the Arsenal being confined to the instruction of the fourth, or beginning class, the cadets then being transferred to the Citadel for the last three years' course.

While the conditions for the oversight and control of the cadets at the Citadel were ideal so far as the physical plant was concerned, it need not be supposed that the administration of discipline could be altogether secured by such means, and that many cases would not arise calling for tact as well as firmness. No doubt the authorities would have been glad at times to relax the rigidity of the regulations if only the offenders would have somewhat relaxed in their attitude of rebellion. An instance in illustration is recorded in the report of the Board to the legislature for the year 1858:

"The Board regret to state that the usual high discipline at the Citadel Academy could only be maintained by the dismissal of a large portion of the Second (Junior) Class, who thought that they had the right to construe the regulations of the institution according to the meaning which they supposed reasonable or convenient; and *as they persisted in adhering to that determination,* the Board had no other course left than to confirm the action which had been taken in their case, which was as follows: During the month of February last a section of the Second Class consisting of eleven complained of an order of one of the professors which required them to rise and march out of the section room in a more orderly manner than before, and as their remonstrance against the innovation which was made to the professor through their squad marcher was unheeded, they resolved to disobey the order.

"They did not seek redress in the easy mode permitted by the Regulations, by complaining first to the Superintendent, and then appealing to the Board of Visitors, which they could have done through their squad-marcher, as they did do to their professor, without the necessity of formal meeting for the purpose, but they hastened to the conclusion that their complaint would be disregarded both by the Superintendent and the Board, and they left the section room in open defiance of the order and authority of the professor, whereupon these eleven cadets were suspended until the Board could act upon their case.

"Thirteen others of the same class, in order to show their sympathy for their fellows, requested the Superintendent to put them on the same footing, but he refused to do so, as they had yet done nothing to merit the same punishment. Then they resolved to hold a meeting in premeditated

---

the institution," and who would have put up modern dormitories, the new barracks were built exactly on the plan of the old, with the result that the spirit of the Old Citadel took up its abode in the new with scarcely a consciousness of any change.

disobedience of the Regulations of the Institution, and against the positive orders of the Superintendent, and, in consequence, they were also suspended.

"Ten others of the same class prudently forebore to show any sympathy for the movement or take any part in it.

"In this state of things, the matter was submitted to the Board, upon the report of the whole case by the Superintendent, with a statement of the matter by the suspended members of the class.

"In the meantime, one of the suspended cadets, expressing his regret at the course he had pursued, and referring the whole subject to the decision of the Board, was restored to the institution.[2] The twenty-three others, who still persisted in claiming the right to disobey the constituted authorities of the institution, were dismissed."

The year 1854 was signalized by a march of the cadet corps through the upper counties of the State. On May 10, the Citadel cadets entrained for Columbia, where the Arsenal contingent joined with them in their friendly invasion of the upcountry.

A two-day march brought the young soldiers to Winnsborough, where, despite blistered feet, they gave a parade for the townfolk and spent an enjoyable day. From here the march was through Chester to Yorkville, where the hospitable people spread a great feast for the hungry boys. A delightful dance was also arranged for the cadets according to the record, and there was a local college where the girls were as pretty as the wonderful roses of York; whether or not the girls were at the dance is not clearly set forth. At Limestone Springs, however, where there was another flourishing female college, Dr. Curtis, the president, "let down the bars," as he himself expressed it, and let the girls entertain the young soldiers.

From there the march went on to Spartanburg, Greenville, Laurens, Newberry, where the battalion fired a salute over the grave of Captain W. F. Graham, the first Citadel superintendent, and finally back to the capital city, "everywhere picnics, dances and sumptuous dinners," so that the boys remembered these days all their lives. On leaving each town, Mitchell's colored band, which accompanied the cadets, played "The Girl I Left Behind Me" as a compliment from the departing guests.

Thirty-five years later, the idea of introducing the Citadel cadets to the people of the upcountry by holding summer encampments in that portion of the State, was revived, and extended even to include the holding of graduating exercises away from Charleston, the seat of the institution, something unique in college commencements.

---

[2]This young man, F. De Caradeuc, graduated in 1860, entered the army of the Confederacy in 1861, and in 1862 gave his life for his country on the battlefield in Virginia.

On February 22, 1857, on the occasion of the semicentennial celebration of the Washington Light Infantry, of Charleston, the Citadel was honored, and the friendly attachment of the members of that company for the cadets signalized, by the presentation of a standard of colors to the cadet corps:

"Assembling in Meeting Street, in front of the Charleston Hotel, in large force at the hour appointed, the Washington Light Infantry were received by their escort which comprised the Citadel cadets, Major Capers; the Charleston Riflemen, Captain Jos. Johnson; the German Riflemen, Captain Siegling; the Moultrie Guards, Captain Hanckel; and the Palmetto Guards, Captain J. J. Lucas.

"The procession then marched through a gauntlet of admiring spectators, who occupied every point and post of observation, to the Institute Hall."

Governor Alston was present, and was greeted with cheers. A poem on "Washington Day" was read by Miss M. E. Lee; the semicentennial oration was delivered by Honorable W. D. Porter, ex-captain of the company; and the "Banner Song" was sung by the Washington Light Infantry choir.

"All eyes were then interested and attracted by the ceremony of presenting a standard to the Citadel Cadets. This standard is of rare elegance and richness in material and execution, doing credit to the taste of the committee that ordered it, and to the spirit of courteous recognition and fraternity displayed towards the gallant young pupils of the State Military School."

On behalf of the centennial committee, W. S. Elliott, Esquire, presented the standard to Captain Hatch, of the Washington Light Infantry, in an elegant address, concluding with these words: "I would beg, Sir, that you would offer this flag to the Citadel Cadets as a mark of the high esteem and regard of the Washington Light Infantry, feeling assured of its safety and honor among hearts so pure, chivalric, and loyal."

Captain Hatch made a beautiful little address in presenting the flag to Major F. W. Capers, superintendent of the Citadel, who replied fittingly, and then gave the standard into the custody of Cadet Captain W. J. Davis, who responded in fine spirit for the cadets.

Something more will be said later of this standard, which was the battle flag of the corps in the closing days of the War. When the Citadel was reopened in 1882, it was borne for many years by the color guard as the battalion colors alongside the national flag on parades. Today, it is preserved between plates of glass among the treasured relics of the institution.

The Association of Graduates

The Association of Graduates, which was destined to play a notable part in the future history of the Citadel, was organized on Friday, November 19, 1852. The first notice of the meeting appeared as an advertisement in the *Courier* of Wednesday, November 17, as follows:

CITADEL ACADEMY

"The graduates of the above Institution are requested to meet at the Citadel on Friday evening at 7 o'clock in order to organize a Society of the Alumni."

The meeting was held at the appointed time, and an organization effected with C. C. Tew as president and J. P. Thomas as secretary, with other officers.

Next year, on Friday, November 25, 1853, the first anniversary celebration was held with an elaborate banquet at the new Mills House. Nickerson's bill of fare was a marvel of superabundance, and deserves to be recorded as the acme of culinary service.[3]

```
                BILL OF FARE
               Soup—Oyster
                   Fish
         Fresh Codfish—Egg Sauce
                  Boiled
         Leg of Mutton—Caper Sauce
         Tongue, Turkey, Celery Sauce
                 Entrées
         Filet of Beef with Truffles
            Chicken Croquettes
                 Macaroni
       Filet of Pheasant with Mushrooms
         Mutton Chops, Sauce Surbese
               Fried Oysters
                Vegetables
    Boiled, Baked, and Roasted Irish Potatoes
     Rice, Beets, Fried Parsnips, Green Peas,
             Beans, Onions, Spinach
                  Roasts
           Beef-Goose, Apple Sauce
               Capons, Turkeys,
            Ham, Champagne Sauce
                   Game
            Canvas Back Ducks
        Pheasants, Venison, Wild Turkey
                Pastry, etc.
      Rusk Pudding, Apple Pie, Cranberry Pie
    Louisene Cakes, Omelet Soufflés, Lady Fingers
    President Meringues, Bavarian Cheese, Mince Jelly
                  Dessert
          Fruits, etc. Coffee and Liqueurs
```

---

[3]A similar elaborate menu of the New England Society is recorded in the history of that society by Dr. William Way, president, for the annual dinner, December 22, 1856.

The wine list was not given, but after thirteen regular toasts were drunk, volunteer toasts became the order of the night, someone proposing a sentiment and calling on another individual to respond. This speaker, after making response, would close by announcing a new sentiment and calling on someone else in turn, until everybody, according to the record, had spoken at least once.

The reporter for the *Courier* stood gallantly by his job for a considerable time, but finally succumbed. In a later edition of the paper he explained: "Other toasts were given which we were able neither to *collect* or *recollect*," and gives due credit to the contemporary newspaper as follows: "We cull from the columns of our neighbor, *The Standard*, the following addenda to our account of the anniversary supper of this association at the Mills House on Friday evening last." Apparently, the reporter for the *Standard* could stand a little more than the *Courier's* man.

Among the guests at this great banquet were some very prominent men, officials of the State. Some notable speeches were made, such as those of General Jamison, ex-Governor Means, Honorable R. F. W. Allston, Richard Yeadon, and others. It appears that this was the only banquet given by the ante bellum association. At least, according to the public prints, the subsequent meetings were distinguished only by public oratory.

At the annual meeting in 1853, J. L. Branch was elected president and P. F. Stevens secretary, to hold office for two years. In 1855, Mr. Branch was reëlected president for two more years, with J. J. Lucas as secretary. In the 1857 election Branch and Lucas were retained in office.

It was customary at the annual meetings for the association to be addressed by some specially invited speaker. The orators on the several occasions were: 1853—Colonel Johnson Hagood, of Barnwell. 1854—Professor C. C. Tew, of the Citadel. 1855—F. F. Warley, Esquire. 1856—Professor P. F. Stevens. 1857—W. J. Magill, Esquire. 1858—None; Commencement changed from November to April. 1859—Professor J. P. Thomas. 1860—Captain Asbury Coward.[4]

Just before the Commencement in 1861, the bombardment of Fort Sumter took place, and the State was at war. There were no formal exercises held, and the graduates went immediately into service. For the next four years, the institution continued to turn out its graduates for the service of the State, but there were no alumni meetings at the Citadel, because the graduates were in the field.

---

[4]Half a century afterwards—June 30, 1909—Colonel Coward, who had the year before retired as superintendent of the Citadel after eighteen years' service, was invited to make the Commencement address to the graduating class. In the pavilion at the Isle of Palms, he repeated his address of 1860—a unique circumstance.

The Irrepressible Conflict

On April 17, 1860, the Democratic National Convention met in Charleston to select candidates for President and Vice-President.

Here occurred the first fatal secession. The convention met in Institute Hall, where, later in the year, the ordinance of secession was signed.

Yancey, the delegate from Alabama, declared in the convention that there was a widening breach between the Democrats of the North and the South, and with impassioned oratory uttered the prophetic words: "Defeat is better than victory won at the expense of principle." When the convention split into two irreconcilable groups which later met separately at Richmond and Baltimore, to nominate rival tickets, the defeat of the divided Democracy was assured.

The election of Lincoln and the triumph of the young but aggressive Republican party made immediately inevitable the conflict between the opposing political ideas of State rights and Federal power, which had been made more bitter by the implacable fight of the Abolitionists against the institution of slavery. The long struggle between the sections, which had been carried on up to that time in the political forum, was now to be decided on the field of battle by the arbitrament of the sword.

In the very beginning of this conflict, the Citadel was destined to play a part.

That a crisis in the affairs of the country was at hand was generally recognized North and South. South Carolina took the lead in acting.

A convention to consider secession was called, and met in Columbia on December 17, 1860. It happened that smallpox had broken out in the town at the time, so the convention adjourned to Charleston. And thus it was that this city, already rich in historic annals was to have, by chance, another distinction added to it as "the cradle of secession."

The president of the convention was General David F. Jamison, author of the bill to establish the Citadel in 1842, and one of the pillars of the institution up to the time of his death.

The convention delegates were received in Charleston with enthusiasm. The battalion of Citadel Cadets paraded in their honor, and a salute of fifteen guns was fired by the Washington Artillery, under the command, appropriately enough, of Lieutenant Salvo.

The convention met first in Institute Hall, near the Circular Church on Meeting Street, but changed next day to specially prepared quarters in St. Andrew's Hall, near the Catholic Cathedral of St. Finbar, on Broad Street; and it was in this hall that the fateful Ordinance of Secession was unanimously adopted at 1:15 P.M., December 20, 1860. The signing of the ordinance was made a formal occasion on the evening of this day, when the delegates reassembled at St. Andrew's Hall at 6:30 o'clock and formed a procession which marched to Institute Hall, where the delegates subscribed their names to the momentous document.

The incident has often been recited that James L. Petigru, the eminent lawyer and Unionist, when he heard the ringing of the bells after the passing of the Ordinance, inquired where the fire was, and was told that there was no fire, but that the city was celebrating secession. He replied: "I tell you there is a fire: they have this day set a blazing torch to the temple of constitutional liberty, and we shall have no more peace."

By a remarkable fatality, a year later, on December 11, 1861, a great fire swept across the city from East Bay southwestward, destroying six hundred buildings, among them the Circular Church and Institute Hall on Meeting Street, and further on St. Andrew's Society Hall and St. Finbar's Cathedral, and a little later the home of Mr. Petigru at 107 Broad Street.

The table on which the Ordinance had been signed, and which had been presented to the St. Andrew's Society by the secession convention, was saved along with some other property, and is still one of the cherished relics of this ancient Scottish society.

General Robert E. Lee, who, on November 8, 1861, had been sent to take command of the department of South Carolina and Georgia, witnessed the conflagration with members of his staff from the roof of the Mills House on the corner of Meeting and Queen Streets, until the approaching fire forced them to seek safer quarters elsewhere.

Although South Carolina had resumed her sovereignty as an independent State, it was hoped that no armed conflict with the United States would result. Governor Pickens, however, immediately took steps to organize the military forces of the State for any emergency.

A critical situation existed in Charleston, the metropolis of the State: Fort Moultrie, on Sullivan's Island, and Fort Sumter, at the entrance to the harbor, were both occupied by United States troops. At the United States arsenal in the city was a bare guard, which was relieved by the Washington Light Infantry even before the signing of the Ordinance. It was necessary that these properties should be in the hands of the State, and it was this delicate situation that ultimately precipitated hostilities.

In the interim between the election of Lincoln and his inauguration four months later, South Carolina made efforts to come to an understanding with the government at Washington about the release of the forts to the State, but without success.

Major Robert Anderson was in command of the United States troops, a battalion of regular artillery, at Fort Moultrie. The State's military forces

The "Star of the West"

in Charleston consisted of the Fourth Brigade, made up of two regiments of infantry, the Rifle Regiment under Colonel Johnston Pettigrew, the Seventeenth Regiment under Colonel Cunningham, four batteries of light artillery, the Washington, Lafayette, Marion, and German; and two troops of cavalry, the Charleston Light Dragoons, and the German Hussars. The battalion of Citadel Cadets, under Major P. F. Stevens, was also counted among the military commands available for service.

On the night of December 26, Major Anderson secretly evacuated Fort Moultrie and withdrew his small force to Fort Sumter, spiking the Moultrie guns behind him. This was construed by the State as tantamount to an act of war. Colonel Pettigrew and Major Ellison Capers were sent by Governor Pickens to get an explanation. Major Anderson's only reply was that as both forts were in his military district, he was at liberty as commanding officer to occupy either post he pleased.

In the meantime, the State forces promptly occupied Fort Moultrie; and a detachment of the Rifle Regiment under Colonel Pettigrew and Major Capers, who was at this time assistant professor at the Citadel, occupied Castle Pinckney in the inner bay. In order to protect the entrance to the harbor, Morris Island was selected for such hasty fortifications as could be constructed of sand, and on New Year's Day of the fateful year, 1861, the Zouave Cadets, German Riflemen, and Citadel Cadets under the command of Lieutenant Colonel John L. Branch, of the First Regiment of Rifles, and a Citadel graduate of the Class of 1846, occupied Cummings Point on this island, and began the construction of the sand battery. When completed, it was armed with four smooth-bore, 24-pounder cannon, and manned by the Citadel Cadets, who were more familiar with artillery drill than the other commands.

And now the scene was set for the first act of the tragic drama of the War Between the States.

News had been received of the fitting out of the *Star of the West,* a steamer with provisions, munitions, and a reënforcement of two hundred and fifty men for Major Anderson at Fort Sumter, and the little battery at Cummings Point was charged with the duty of seeing that they should not pass. The guard boat was notified of the expected coming of the steamer, and sentinels kept watch along the beach of Morris Island.

At daybreak on January 9, 1861, the guard boat sighted the *Star of the West,* and gave the signal. What then took place at the sand battery is thus described by Sergeant S. E. Welch, of the Zouave Cadets, who was on duty at the time:

"Cadet W. S. Simkins, on post on the Battery, gave the alarm, the sentinels along the beach took up the call, the long roll was sounded, and the men immediately took their positions, the Citadel Cadets at the guns, the Zouave Cadets and German Riflemen just in their rear as an infantry support.

"The Ship was soon inside the channel and rapidly approaching. The guns were loaded, the lanyards stretched, the men awaiting orders. There seemed to be some hesitation among the higher officers; the commanding officer, evidently impressed with the seriousness of firing on the United States flag, appeared to be in doubt just what to do.

"Major P. F. Stevens, commanding the Cadets, turned and gave the command: 'Commence Firing.' The cadet captain passed the order: 'Number One, Fire!' Cadet G. E. Haynesworth of Sumter, pulled the lanyard and fired the first gun of the War, the shot going across the *Star of the West*.

"Cadet S. B. Pickens fired the second shot, directly at her, and the firing then became general, each gun in turn. The vessel paid no attention to the first shots; then slowed down; turned, and put out to sea."

The report of the action made by Captain McGown, who commanded the *Star of the West,* is as follows:

"When we arrived about two miles from Fort Moultrie—Fort Sumter being about the same distance—a masked battery on Morris Island where there was a red palmetto flag flying, opened fire upon us—distance about five-eighths of a mile. We had the American flag flying at our flagstaff at that time, and soon after the first shot hoisted a large American ensign at the fore. We continued on under the fire of the battery for over ten minutes, several of the shots going clean over us. One passed just clear of the pilot house. Another passed between the smoke-stack and the walking-beam of the engine. Another struck the ship just abaft the fore-rigging, and stove in the planking, and another came within an ace of carrying away the rudder."

The Charleston *Mercury* of January 15 gave the following extracts from the New York *Evening Post,* which published a full account of the voyage of the *Star of the West:*

"The military men on board highly complimented the South Carolinians on their shooting in this first attempt. They say it was well done; that all that was needed was a little better range, which they probably could have obtained in a few minutes. Their line was perfect; and the opinion is expressed that some one had charge of the guns who knew his business. . . . Two guns were employed, the smaller, it is believed, a twelve-pounder, and the larger a thirty-two pounder. This however, is only conjecture. Whatever their size, they were well manned. They were fired rapidly and with will.

"One of the officers hazarded a joke soon after we left Charleston harbor; 'The people of Charleston,' he remarked, 'pride themselves upon their hospitality, but it exceeds my expectation—they gave us several *balls* before we landed!' "

THE WAR PERIOD

The Charleston *Mercury* of January 10, 1861, in its account of the *Star of the West* incident, used the headlines:

THE WAR BEGUN.
ENGAGEMENT AT FORT MORRIS.
ATTEMPTED REINFORCEMENT OF FORT SUMTER.
THE "STAR OF THE WEST" IS FIRED INTO
AND DRIVEN BACK.
THE CITADEL CADETS FIRE THE FIRST SHOTTED GUN.
THE UNITED STATES FLAG HAULED DOWN.
THREE OF THE SHOTS TAKE EFFECT.

Similarly, *Harper's Weekly,* New York, in its issue of January 19, 1861, headlines its article on the engagement: "The First of the War" and gives its readers a picture of the *Star of the West,* and a map of Charleston harbor.

But subsequently the beginning of the war was fixed at a later date—April 12, 1861—and the events of January 9 were not charged with the burden of responsibility for the awful conflict.

After the departure of the *Star of the West,* the cadets and other military organizations remained on duty in expectation of a return of the vessel. But diplomacy resumed control of affairs, and in the latter part of January the cadets returned to barracks. The seniors, however, were doing more drilling of recruits than reciting in the classroom, and the Citadel Green was a busy muster ground.

As Commencement in April approached, it became apparent that a climax in the affairs of the State was imminent.

South Carolina had been joined by five sister States in February, and the government of The Confederate States of America had been organized with its seat at Montgomery, Alabama. Charleston was recognized as the crater where the eruption would take place. When the Board of Visitors of the Citadel met on April 9, preparatory to the annual Commencement, the rumbling of the volcano was so threatening that the Board released the members of the graduating class, awarding them their diplomas without the usual formalities. The following resolution was passed: "That in consequence of the imminent collision between the troops of the Confederate States and the forces of the United States in the immediate vicinity of the city of Charleston, the usual ceremonies of the Commencement be dispensed with."

The Board, consisting of General James Jones, General D. F. Jamison, General J. H. Means, Colonel I. D. Wilson, Colonel Henry C. Young, and the Adjutant General, S. R. Gist, probably witnessed the bombardment of Fort Sumter on April 12, as they were in Charleston on the 15th, on which date they held another meeting to transact Citadel business. Among the matters acted on at this meeting was the dismissal of six cadets who had been suspended for creating a disturbance at the Citadel. It is worthy of note that in taking this action, the Board "highly commended the efficiency and soldierly bearing" of one of these suspended cadets "during the affair of Fort Sumter," but stated that they could not "endanger the discipline of the Academy by restoring him to his cadetship." This incident is illustrative not only of the difficulty of preserving discipline in those unsettled times, but also of the firmness of the Board in dealing with cases of insubordination.

In the memorable bombardment of Fort Sumter, the services of three Citadel officers are mentioned in orders by General Beauregard:

"To Major Stevens, of the Military Academy, in charge of the Cummings Point batteries, I feel much indebted for his valuable and scientific assistance, and the efficient working of the batteries under his immediate charge. . . . I would also mention in terms of praise the following commanders of batteries at the Point, viz.: Lieutenant Armstrong, of the Citadel Academy; . . . also, Captain Thomas, of the Citadel Academy, who had charge of the rifled cannon, and had the honor of using this valuable weapon, a gift of one of South Carolina's sons to his native State, with singular effect."

The Citadel also promptly undertook a military service of another kind for the State. The first laboratory for the manufacture of ordnance stores to be established in the Confederacy was organized at the Academy, and rendered valuable aid during the early days of the War.[5]

But the call to the battle line seemed to many of the Citadel officers more urgent than any service, however important, which they might render at the Citadel. On August 8, 1861, Major Stevens, the superintendent, resigned, and went to Virginia as colonel of the Holcombe Legion, taking part in many engagements, and being wounded in the battle of Sharpsburg. Another member of the faculty, Lieutenant Ellison Capers, resigned

---

[5] In June, 1861, Professor P. F. Stevens and E. Capers, of the Citadel Academy, tested out a 24-pound smooth-bore cannon which had been rifled. They took the gun up the South Carolina Railroad near Summerville, and fired about 100 shots with it before it burst. They reported that if bands should be shrunk around the breech, the largest guns in the harbor might be rifled and used effectively.

November 29, to enter the Confederate army in the West, and made a distinguished record, rising to the rank of Brigadier General. The two honor graduates of the Class of 1861, C. I. Walker and J. D. Lee, were elected by the Board as assistant professors at the Citadel, but declined the appointments, feeling the call to the colors the higher duty. Both rendered conspicuous service in the Confederate Army, Walker as lieutenant colonel of the 10th South Carolina Regiment in the Western Army, and Lee in Virginia, where he was killed in battle near Richmond.

CHAPTER V
FIFTH DECADE: 1862-1872
War Times

IT may easily be surmised that the educational institutions of the State suffered greatly in attendance during the period of the War, since many of the young men who would normally be in college were serving in the army. The depletion occurred first among the older students, juniors and seniors, but soon extended to all the classes. The flower of the young manhood of the country was, indeed, engaged, not with books, but in battle with the enemy.

Before the end of the second year of the War, the student bodies at some of the colleges were scarcely larger than the faculties. At the beginning of 1862, there were seventy-two students at the South Carolina College, but on March 8, when the orders of council were published, bringing within the provision of the conscription practically the entire student body, the boys held a meeting and resolved to withdraw from the college and enter the volunteer service. This reduced the number of students to nine —four sophomores and five freshmen.

The denominational colleges, Wofford, Furman, Erskine, and Newberry, suffered in the same way, if not to the same degree. The boards of trustees tried to keep the college plants and faculties intact, but mills and millers are both useless when there is no grist to grind. In the summer of 1862, most of the buildings of the South Carolina College in Columbia were turned over to the Confederate authorities for the more urgent need of hospital facilities, and in the summer and fall of that year more than two thousand sick and disabled soldiers were cared for on the campus. The college remained closed until after the war.

But the reason for the reduced attendance at the "civilian" colleges would not apply with the same force to military institutions like the Citadel. It is true that as the war wore on its duration became more and more a matter of doubt, and a boy about to enter college would not be at all certain that his course would not be curtailed by a call to service. But in this case the advantage of a course at a military school was apparent: he would not only be receiving instruction in the arts and sciences generally, but in the art and science of war in particular. In addition to this, when he matriculated at the Citadel, he would be immediately in the military service, subject to the call of the commander-in-chief. That these considerations appealed to the young men of the State is attested by the greatly increased number of applications for admission to the military academies.

The Board of Visitors, in its annual report to the General Assembly December 1, 1862, stated that a larger number of applications had been received than ever before, and that for want of accommodations they had been forced to reject more than two hundred applications for admission to the class beginning January 1, 1863. They suggested that if the legislature would place at their disposal "a portion of the College buildings, now unoccupied, they would endeavor to make arrangements to receive all who have applied." This, however, was not done.

Two years later, December 2, 1864, the Board considered applications from 449 pay cadets and eighty-seven State cadets, "of whom the Board appointed 190, which, with seven appointed by the Governor under a resolution of the General Assembly, make the whole number of appointments one hundred and ninety-seven. The Board regrets to have to say that the impracticability of erecting, or hiring, or buying buildings for their accommodation precluded any more appointments."

On January 1, 1862, the enrollment at the Citadel was 133 cadets, and at the Arsenal, 139; on the same date in 1863, the Citadel had enrolled 121, the Arsenal 147; and in 1864, the Citadel had 145, and the Arsenal, 151. The estimate of the Board for January 1, 1865, which was made at the last meeting of the body on December 2, 1864, gave the enrollment at the Citadel as 205, and at the Arsenal, 157.

The Battalion of State Cadets

The use of the cadets of the Citadel and Arsenal academies for military purposes was authorized by an act of the legislature passed January 28, 1861, which provided that the two bodies of students should be organized as "The Battalion of State Cadets."

Following is the full text of the act:

"I. That the Arsenal Academy and the Citadel Academy shall retain the same distinctive titles, but they shall together constitute and be entitled 'The South Carolina Military Academy.'

"II. That the officers and students thereof, organized as 'a Public Guard' into one or more companies at each Academy, shall constitute a military corps entitled 'The Battalion of State Cadets.' That said battalion shall be part of the military organization of the State, under the separate and immediate control of the Board of Visitors, and shall not be subject to the command of the militia officers, except when specially ordered for parade, review, or service by the commander-in-chief; that the officers of said battalion shall be commissioned by the Governor, with such rank and titles (the highest not exceeding that of Major) as the Board of Visitors may determine; provided, nevertheless, that the officers of said battalion may be removed by the Board of Visitors, and their commissions thereby vacated, in like manner as is now provided for in the second section of said Act for the removal of professors of the Academy; that the said Battalion of State Cadets, while habitually maneuvering as infantry, may yet maneuver in any arm of the service, and it shall take the right of all troops of the same arm in which it may at any time parade.

"III. That all graduates of the South Carolina Military Academy, in consideration of their four years' service at the said Academy shall be eligible to any commissioned office, not above the grade of colonel, in the military organization of the State."

This act is remarkable in that it makes the student body of a State educational institution a part of the military forces of the State.

It has been seen that Governor Pickens did not hesitate to call on the superintendent and cadets of the Citadel for military service when the *Star of the West* was sent to relieve Fort Sumter. If any doubt existed as to the authority of the governor to make such use of the students of the military academies the matter was settled for the future by this act.

Over half a century afterwards, during the World War, when the colleges of the country were threatened with a paralysis of their operations by the Draft Act of 1917, a similar device was adopted nationally, with the result that instead of empty halls of learning, the dormitories and classrooms of the educational institutions were crowded as never before. Fortunately, the experiment of the students' army training corps did not have to be continued long enough to produce any ineffaceable impressions on the colleges where they were organized.

Prices During the War

At the beginning of the War in 1861, the fees charged pay cadets for board, uniforms, and tuition amounted to only $200 a year. On December 1, 1862, these fees were raised to $400, and certain items previously included in the fees were charged for extra. The estimate of the cost of supporting a cadet at the Citadel at this time was $515.66, more than half of which was for his clothing, $271.25.

On November 28, 1863, the rates were again doubled, pay cadets being required to pay $800, and also "to furnish their own underclothing, including shirts, drawers, and socks; also their shoes, combs and brushes."

During the year 1864, "The Academy was allowed the privilege of purchasing subsistence to a considerable extent from the State and Confederate commissioners at cost price, greatly below the price for which similar supplies were selling in the open market." On December 3, of that year, however, the Board reported to the legislature:

"The Board regret to state that this source of supplies is now cut off, and both subsistence and clothing will have to be purchased at the enormous prices asked in the open market. The Board has therefore increased the amount to be paid by Pay Cadets to twelve hundred for the year, and they ask of your honorable body a corresponding increase in the appropriation for the Academy for the pay of the officers and support of State Cadets."

Also, "in answer to the earnest petition of the officers of the institution, and in pursuance of what seemed to the Board an act of justice to them, their pay was increased one hundred per cent upon their salaries of last year."

These provisions, however, never went into effect.

In the same report, the Board states: "The continued bombardment of the city of Charleston by the enemy has rendered the Citadel in that city a dangerous habitation and wholly unfit for the academic duties of the cadets. They must, therefore, be removed from that vicinage. If convenient and suitable quarters cannot be hired for them, it is the design of the Board to erect temporary huts for their accommodation."

The two battalions were then in the field, and the Citadel cadets never returned to their barracks.

The last financial statement of the Board for the fiscal year ending September 30, 1864, is interesting:

Receipts:
   Balance on hand, Oct. 1, 1863 .... $ 245.59
   Received from appropriations and deposits 113,571.57
   Received from pay cadets 113,200.00
   Received from sales to officers, etc. 8,431.37

                                                                        $235,448.53

Expenditures:
   For pay of officers .... $ 22,003.33
   For subsistence 108,415.26
   For clothing 53,877.26
   For washing 7,387.24
   For hire of servants 2,554.19
   For music 258.00
   For fuel, lights, books, stationery, medicines, etc. ... 24,388.01
   For permanent improvements 76.25
   For repairs to buildings, furniture, etc. 1,348.65
   For Magazine Guard in Charleston 4,926.36
   Amount redeposited in Treasury 200.00
      To balance 10,013.98

                                                                         $235,448.53

At the beginning of the academic (and calendar) year 1864, the enrollment at the Citadel was recorded as 145, and at the Arsenal as 151. In November there were probably less than a hundred in each institution.

Died for Their Country

It would be supposed that the graduates of the Citadel—men who had been trained to the use of arms for the service of their State—would answer the call of their country when it came. Statistics show that of the 224 alumni of the classes from 1846 to 1864 inclusive, living at the time of the war, 193, or 86 per cent, served in the armies of the Confederacy.

Some of the young men who had graduated from the Academy had wandered into distant States and become established in business; but, wherever they were, they answered promptly to the call to the colors. Some of them rose to high command, and all, so far as the records show, met their duties and responsibilities as worthy sons of their Alma Mater.

This is the roll:

1 Major General.
3 Brigadier Generals.
17 Colonels.
10 Lieutenant Colonels.
22 Majors.
58 Captains.
62 Lieutenants.
20 Not commissioned.

Thirty-nine were killed in battle, and four others died in service. Twenty-nine others were wounded, some seriously, and many, more than once.

It is not designed in this brief story of the institution to give the record of their achievements; but no account of the Academy could leave out some reference to their exploits on the field of battle, or fail to call the roll, at least, of those who made the supreme sacrifice for their country. Colonel Thomas's history of the Academy is a thesaurus of great value to those interested in knowing something of the War-time heroes of the Citadel, and the present writer has gleaned most of his information about them from this invaluable volume.

The first major test of strength between the Federal and Confederate forces occurred at Bull Run, near Manassas Station, in Virginia, where McDowell's well-equipped and confident army met Beauregard's and Johnston's Southern troops. It was here, on Sunday, July 21, 1861, that the first Citadel graduate fell in battle, R. A. Palmer, of the Class of 1852. Captain A. D. Hoke, '54, 2nd South Carolina, and Lieutenant Colonel E.

M. Law, '56, of the 4th Alabama Regiment, who afterwards rose to the rank of major general, were severely wounded at the same time.

The rout of the Union Army, and the stampede of the civilian onlookers who had journeyed out from Washington to see the rebels dispersed, made the North realize that the War was a serious enterprise, and the second half of the year 1861 was signalized by large preparations for important operations in the spring of 1862. These developed simultaneously in two great movements—one by McClellan in the Peninsula, with Richmond, the Confederate capital, as its objective, and the other in the middle West under Grant, whose purpose was to open up and control the Mississippi River.

Early in April, 1862, the desperate and bloody battle of Shiloh was fought in West Tennessee between Grant and General Albert Sidney Johnston. The result was altogether favorable to the Confederates on the first day, April 6th; but Johnston was killed, and on the next day, Grant having been reënforced by Buell, the tide of battle turned in favor of the Union forces.

In this terrible battle, John Mills Dean, '55, lost his life at the age of twenty-seven. Having moved to Arkansas prior to the War, he had entered the Confederate Army as lieutenant colonel of the 7th Arkansas Infantry. All day Sunday, the first day of the battle, Colonel Dean's command was in the thick of the fight. It was about five o'clock in the afternoon, while leading a victorious charge in the famous Hornet's Nest, that he was shot through the neck, dying instantly.

In this year, death was to take toll of more than a dozen other lives of Citadel graduates on the battlefields of Virginia.

On the 31st day of May, Lieutenant T. M. Wylie, '61, of the 6th South Carolina Volunteers, received a wound at Seven Pines, which later caused his death. Captain J. W. Daniels, '52, was also severely wounded in this engagement.

At Frazier's Farm, on June 30, another young graduate, barely twenty-one years of age, gave up his life for his country. This was John Dozier Lee, '61, adjutant of Colonel Micah Jenkins's Palmetto Sharpshooters, killed at the head of his regiment leading a charge against the enemy, both legs being taken off by a cannon ball.

At the important battle of Second Manassas, August 29-30, where Jackson so signally outmaneuvered Pope, Major F. Gendron Palmer, '51, was mortally wounded in the right breast, and Major J. M. Whilden, '61, another mere lad, was killed while waving the flag at the head of his regiment, the 23d South Carolina Infantry. J. A. Finch, '56, of the 6th South Carolina Regiment, also fell on this victorious field, and Lieutenant Colonel

W. J. Crawley, '55, of the Holcombe Legion, was severely wounded. It was at this battle, also that ex-Governor Means, chivalrous soldier and one of the great "triumvirate" of the Citadel Board of Visitors, was mortally wounded.

During this year, T. O. McCaslin, '59, was killed in Virginia, and G. Ross, captain of a company of Arkansas Volunteers, died in service, as did also F. de Caradeuc, '60, and R. Noble, '61.

September 13, at Maryland Heights, Captain W. E. Cothran, '59, was severely wounded, and his classmate, J. L. Litchfield, was mortally wounded, dying the day before the sanguinary battle of Sharpsburg, where Colonel C. C. Tew, "proto-graduate" of the Citadel, '46, and N. Wilson, recent graduate, '61, both lost their lives. Two other graduates, W. J. Magill, '46, and Colonel P. F. Stevens, '49, were seriously wounded in this stubbornly contested battle, the former losing his left arm at the shoulder. Cadet J. K. Law, a Citadel cadet on furlough, was acting as a volunteer aide to General E. M. Law in this battle, and was severely wounded, recovering, however, and graduating next year in the Class of 1863.

The death of Colonel Tew was peculiarly tragic. In company with General John B. Gordon, he was reconnoitering the enemy lines near Bloody Lane, when he was struck in the head, and fell, supposedly dead. The Confederates were in retreat at the moment, and his body was never recovered—a circumstance which gave rise after the War to stories of his being alive and held a prisoner. It was not until 1885 that an end was put to these rumors, so cruel to the relatives. In this year, a former Union soldier wrote to a daughter of Colonel Tew that he had seen the Colonel on the battlefield, mortally wounded by a ball which had passed through his temples, dislodging the eyes from their sockets; that he was sitting up with his back against the bank of a road, perhaps unconscious, but still holding his sword. This sword, which was taken from him by Union soldiers, was inscribed, according to the informant, with the words: "Presented to Captain C. C. Tew by the Arsenal Cadets of South Carolina."

Finally, in the last month of this year, in the important battle of Fredericksburg, two more Citadel graduates made the supreme sacrifice, Captain W. S. Brewster, '52, killed in battle, and W. F. McKewn, who had just graduated in the preceding April, mortally wounded.

In the year 1863, at the great Confederate victory of Chancellorsville, May 2-4, Captain T. W. Fitzgerald, '52, of the 12th Alabama Volunteers, was mortally wounded, and two months later at Gettysburg three more graduates fell on that fiercely contested field—D. T. Williams, '52, G. M. McDowell, '59, and Major J. H. Burns, '61. The last-named had participated in First Bull Run, Seven Pines, Second Bull Run, and Fredericksburg; had been wounded at Sharpsburg, and now laid down his life in the decisive battle of the War. Colonel Edward Croft, '56, of the 14th South Carolina Volunteers, and Captain F. M. Farr, who had graduated just a few months before, were both severely wounded in this battle.

In this year, also, J. B. Colding, '51, was killed at Winchester, Virginia, E. A. Erwin, '60, on Sullivan's Island, and F. H. Harleston, first-honor graduate of the Class of 1860, fell on the ramparts of Fort Sumter.

In the great battle of Chickamauga, on September 20, 1863, Lieutenant Joseph C. Palmer, Adjutant of the 24th South Carolina Volunteers, while riding with Lieutenant Colonel Ellison Capers, leading the charge of the regiment, was struck full in the forehead by a ball, and expired instantly. "No gallanter soldier gave his life for his country."

Chancellorsville had marked the high tide of Confederate success, and at the decisive battle of Gettysburg there began a perceptible recession. Beginning with the year 1864, the South realized that she was fighting a losing battle.

In this fateful year, ten more heroic names were added to the roll of Citadel dead. On May 6, occurred the tragic death of Brigadier General Micah Jenkins, '54. It was in the Battle of the Wilderness, and the details are thus given in Colonel Thomas's history:

"The lamented fall of Micah Jenkins is thus described by that gallant young soldier, Colonel J. R. Hagood: 'I cannot forget the appearance of General Jenkins this morning. Elegantly dressed, as he always was, superbly mounted, and his face lit up with martial fire, he realized to the full my ideal of a soldier. At this moment General Longstreet, accompanied by General Jenkins and a number of staff officers, galloped ahead of the line until they had reached the plank road. There an orderly dismounted to take up a stand of Federal colors lying upon the field, and displaying them, Mahone's Virginia Brigade, advancing through the thick woods mistook the party for a body of the enemy's cavalry, and a volley from them emptied the saddles of Longstreet, Jenkins, a staff officer and two or three orderlies. General Jenkins was instantly killed, and General Longstreet badly wounded.'

"This is how Jenkins died on the second day of the Battle of the Wilderness.

"Were the historian to take up Jenkins' career in the Confederate War and follow him from battlefield to battlefield, and tell of his acts of personal heroism, of his military skill, and daring generalship; were he to relate how he won the confidence and admiration of his military superiors; were he to relate further the wealth of love he bore for family, for friends, and country, and the innate nobility of his nature, the splendid record would occupy a volume in itself.

"Gallant spirit, noble friend! We greet your memory as we place your name high upon the historic page, and renew our lament for your death in the bloom of manhood!"

On the same day on the same battlefield, fell another gallant son of the Citadel, Colonel James D. Nance, '56, colonel of the 3rd South Carolina Volunteers, Kershaw's Brigade, Longstreet's Corps, at the youthful age of twenty-seven, and on the eve of promotion to brigadier general.

"His Country called her Sons to War;
He perished among the Slain in her Defense."

In this month, also, four other Citadel graduates died in battle: J. B. Dotterer, '63, mortally wounded at Resaca; J. A. Evans, '56, at Kennesaw Mountain in North Georgia; O. J. Youmans and Colonel W. P. Shooter, both of the Class of 1859, in Virginia. Then followed, G. B. Dyer, '62, killed near Richmond, and W. Mason Smith, '63, mortally wounded at the sanguinary battle of Cold Harbor on June 3. On John's Island, near Charleston, G. A. McDowell, '62, was killed, and on July 30, near Petersburg, Virginia, J. D. Quattlebaum, who had just graduated from the Citadel, and Colonel D. G. Fleming, '54, perished in the Crater. The latter had served many months in the heroic defense of Fort Sumter, and had been transferred to Virginia in the spring of 1864. About daylight on the morning of July 30, he was inspecting the lines with his adjutant and orderly, when the fearful explosion planned and executed by the besieging Federals occurred, and he was buried with many others under the mountainous eruption.

When the year 1865 dawned, the end of the Confederacy was in sight. Death's toll, however, was to include four more graduates of the Military Academy. In January, Sherman's relentless march northward through South Carolina could not be seriously retarded by any efforts of the Southern forces. An engagement at River's Bridge, on the Salkehatchie River, took, among other lives, that of S. S. Kirby, '60; D. P. Campbell was killed at Pocotaligo. In Virginia, Colonel C. W. McCreary, '57, who had served four years with the Army of Northern Virginia, commanding a regiment at Gettysburg, and severely wounded May 12, 1864, near Richmond, was killed in the engagement at Five Forks, Virginia.

Finally, at Bentonville, North Carolina, A. S. Gaillard, '60, received a wound in battle which ultimately caused his death, and made him the last martyr to the Lost Cause which the Citadel gave for the Southern Confederacy.

MILITARY SERVICES OF THE CADETS IN THE WAR

The question has frequently been asked if a cadet at the Citadel or Arsenal during the War of Secession had any "war record" by virtue of his having been a student in the State's military academy. Such a question asked about a student of any other educational institution in the State would be answered in the negative.

But the act of the legislature passed January 28, 1861, specifically stated that the officers and students of the Citadel and Arsenal "shall constitute a military Corps entitled 'The Battalion of State Cadets.' That said battalion shall be part of the military organization of the State, under the separate and immediate control of the Board of Visitors, and shall not be subject to the command of the militia officers except when specially ordered for parade, review, or service by the commander in chief."

As a matter of fact, the services of the cadets were called for on several occasions, as shown in the following account prepared by ex-Governor Hugh S. Thompson, who was captain of the company of Citadel Cadets during the War. This account taken (condensed) from Colonel Thomas's history:

EXTRACT

Of the service of the Cadets in repulsing the "Star of the West" in January, 1861, and of the officers of the Academy in the reduction of Fort Sumter in April of that year, it is not necessary for me to speak.

In November, 1861, immediately after the fall of Port Royal, the Cadets were ordered to support the Washington Artillery stationed at Wappoo Cut. There was then good reason to believe that gun boats from Port Royal would attempt to reach Charleston through the inland route. The Cadets remained on this duty until it became apparent that there was no further danger to be apprehended from this quarter. Academic duties were resumed and continued until the early part of June, 1862, when General Pemberton ordered Major White to take the Cadets to James Island and mount some heavy guns upon the works there. After the guns were mounted, General Pemberton determined to retain the Cadets there, as they seemed to be in great danger of an attack, the Federal troops at this time being very active in the preliminary movement which culminated in the battle of Secessionville.

Major White, who had obeyed the order of General Pemberton without consulting General Jones, chairman of the Board, because he thought the emergency justified his doing so, then directed me to go to Columbia and lay the whole matter before him. General Jones heard me without remark or comment and after I had completed my statement of the reason which influenced Major White in the course he had taken, wrote a brief and peremptory order to him to remand the corps of Cadets forthwith to the Academy and to report in person to him at Columbia.

I returned immediately to James Island with the order, which I delivered to Major White. He at once proceeded to Charleston and asked General Pemberton to relieve him that he might obey the order of his immediate superior. This General Pemberton positively refused to do; but subsequently, on the advice of Lt.-Col. Wagner, Chief Ordnance Officer, at whose suggestion the Cadets had been ordered to James Island, he relieved them from further duty there.

The course of General Jones in this matter caused much disappointment and even mortification among the officers and cadets, because it was believed that the city was in danger, and that they should have been allowed to remain at the outer defences.

As a consequence of this feeling, thirty-six cadets (five in the First, or graduating, class, eleven in the Second Class, and twenty in the Third Class) deserted and formed a cavalry company[1] which afterwards did gallant service in the war.

Many of the cadets who remained in the Academy were much dissatisfied and wished to withdraw. At the request of Major White, Judge Magrath made an address to them, explaining earnestly and eloquently their duty to the State at this critical moment. The effect of this speech was most beneficial, and the cadets returned to their duties with fresh spirit and energy.

Beyond occasional tours of guard duty in different parts of the city, no further military service was rendered by the Cadets until July 10, 1863, when Federal troops captured the upper end of Morris Island. Early on that day, General Ripley ordered Major White to hold the cadets in readiness for duty, and about noon instructions were received to take the battalion to Morris Island to assist in the defense of the batteries there. The order was promptly obeyed, and soon the Cadets were at the wharf awaiting transportation. Before the boat which was to take them to the Island arrived, however, an order was received from General Ripley directing them to return to their quarters, as their services were not then needed. The Cadets received with great regret the order to return, as they were looking forward eagerly to the opportunity to see active service on Morris Island.

The Citadel Cadets remained on duty in the city, and were joined soon afterwards by the Arsenal Cadets, the battalion thus formed remaining in active service in and about Charleston until the first of October, when after a short furlough, they returned to their respective academies.

Late in May, or early in June, 1864, the Cadets were ordered to James Island to resist a threatened attack there. The Confederate forces in the defenses around Charleston were then very weak, in consequence of the heavy withdrawal of troops to Virginia.

---

[1] "The Cadet Company," of which more anon.

The Cadets remained in the discharge of this duty until the danger which called them out had passed. Beyond occasional shells fired at the picket lines, they were not exposed to the Federal fire, but the performance of picket duty and other service incident to a campaign on James Island in the heat of summer was most trying. But, not only was there no complaint, there was no service that they did not perform cheerfully and in the spirit of true soldiers.

Later in the season furloughs were granted to the Cadets, and when they reported for duty in October yellow fever was prevailing in the city. For this reason, they were ordered into camp near Magnolia, and subsequently at Orangeburg. While at the latter place, in December, 1864, they were ordered to Charleston, and from there to reinforce the troops engaged in defending the line of the Charleston and Savannah Railroad.

Of the incidents of that service it is needless for me to speak to you who were an eyewitness to the gallantry of the Cadets and to the soldierly manner in which they met every requirement. You recollect doubtless, the encomiums they received from Gen. Sam. Jones and others under whom they served, but it may be that you do not recollect what was perhaps the highest compliment ever paid them. The story was related to me by Capt. Bachman soon after the War. You will remember that we served with Bachman's Battery. He said an old soldier who served several years in Virginia expressed his admiration of the battalion of Cadets in the fight at Tulifinny in these words: "Them fellows fight like Hood's Texicans."

After the engagement at Tulifinny on Dec. 6th and 7th, the Battalion went into camp, doing picket duty and being under fire of the enemy's batteries. On Dec. 25th they were withdrawn, the Citadel Cadets going to James Island, where they formed a part of the brigade of Stephen Elliott, and the Arsenal Cadets to Columbia. On the evacuation of Charleston in the middle of February, 1865, the Citadel Cadets retired with Hardee's Army, still under the command of Elliott, who frequently took occasion to express his high admiration for the courage, spirit, and discipline which they showed on that trying march.

*   *   *   *   *

There are two references in the above account by ex-Governor Thompson which deserve amplification. One was to "The Cadet Company," whose exploits are recorded in an article by Lieutenant Alfred Aldrich, an officer of the company. The other was to the "Battle of Tulifinny," which, while a minor engagement, was the one battle in which the battalion of cadets took part.

The Cadet Company
(As told by Lieutenant Alfred Aldrich, of the Company)

In the spring of 1862, the corps of Citadel Cadets was ordered into the field for temporary service in the defense of Charleston, the city being threatened from James Island. This service, though not of long duration had the effect of making the cadets disinclined to resume their studies. After returning to barracks, the dissatisfied cadets decided to leave. Five members of the First, or senior, Class, eleven Second Class men and twenty from the Third Class left the Academy in a body without authority, and, supplemented by some of the Arsenal cadets and recruits from among their friends in the various counties, organized a cavalry company, and joined the Sixth Regiment of South Carolina Cavalry, then organizing under Colonel Hugh Aiken.

This company was known as "The Cadet Company."

The company proved of incalculable assistance to the Sixth Regiment, which was a regiment of recruits, for there was scarcely a cadet of the Citadel who was not competent to drill a company, battalion, or regiment, and the commissioned officers of the Cadet company were detailed by Col. Aiken to form a school of the officers and non-commissioned officers of the regiment and instruct them in Hardee's tactics while his regiment was on duty at Adams Run during the winter of 1863. So well did the cadet officers discharge this duty that Aiken's Regiment of Cavalry, twelve hundred strong, went to Virginia in the spring of 1864 the finest equipped regiment for service that ever left the Palmetto State. And the record will show that no regiment in the Confederate service did more arduous and efficient duty than the Sixth Cavalry, and that no company of the Regiment was so often to the fore as "The Cadet Company."

While still on the Carolina coast, the Cadet Company supported Gen. Del Kemper's Artillery in an attack on the gun-boats, *Pawnee* and *Marblehead,* in Edisto Inlet, on Christmas morning, 1863. A few weeks later, the Company supported the artillery commanded by the gallant Major John Jenkins in an effort to sink the *Pawnee* in Stono River. It also supported Walter's Battery in an artillery battle on John's Island in March, 1864.

On Feb. 9th, the Union forces made an advance in force on John's Island.

The Confederate picket force stationed at "Haulover Cut" was driven in about daylight. Major Jenkins, commanding the advanced forces, ordered Captain Humphrey to take the Cadet Company and Sullivan's Company to ascertain the strength of the force which had driven in our pickets.

Humphrey deployed the battalion as skirmishers and, mounted on his magnificent stallion, *Sago,* moved his command through an open field three hundred yards wide upon the enemy concealed behind a ditch and hedge-row at right angles to "Haulover." When within fifty yards of the hedge-row, the enemy arose and delivered a volley from eleven hundred muskets into our skirmish line. But so spiritedly and steadily did Humphrey's men pour in their fire that when ordered to "fall back" they were not pressed by the enemy, and withdrew without confusion.

The Cadet Company lost nine men killed, wounded, and captured. Sergt. John S. Dutart and private G. A. McDowell were killed instantly. Lieut. Dozier and privates Spann and Mellett were captured. Captain Humphrey's horse was killed under him and he was slightly wounded.

Humphrey's command retreated in the direction of Church Flats until reënforced by the other companies of the Sixth Cavalry and Colquitt's Brigade, when the Confederates made a stand, halted and repulsed the Federals, and before night drove them back over Haulover.

It was shortly after this action that the Sixth Cavalry was ordered to Virginia and assigned to Butler's Division.

Immediately on our arrival in Virginia, we went into action.

Our first service with the Army of Northern Virginia was with the column of cavalry under Lieut. General Hampton, who went in pursuit of Sheridan when he made his last raid into the Valley.

The Confederates overhauled Sheridan at Louisa Court House, on June 11, 1864, and there was a sharp fight between the head of our column and the rear of the enemy's.

The Cadet Company had position near the front. During the fight, Hart's Battery was charged by a column of enemy cavalry when our line was in confusion passing through a railroad excavation. General Hampton dashed up to the right of the Sixth Cavalry to get a force to follow him in a charge to save the battery, and although the confusion was as great at this part of the line as elsewhere, thanks to the Citadel training, Lieutenants Nettles and Aldrich got their company aligned within a few seconds, and Gen. Hampton charged in line of battle with the Cadet Company, staying the enemy's onset until Gen. Rosser struck his flank, charging at the head of his own brigade.

The secret of success in battle, according to the great cavalry leader of the Western Army, Gen. N. B. Forrest, was "to get there first with the most men." Here was a critical moment when the day would have been lost, perhaps, if our great leader had not determined to "get there first" with a handful of boys!

The day succeeding the fight at Louisa Court House, the battle of Trevilians, the most sanguinary cavalry engagement of the war, was fought.

The Cadet Company, whether by design or fortuitously we never knew, was in position where the hottest fighting in this hotly contested engagement occurred.

Our position was in the angle formed by the wagon road crossing the railroad at Trevilians station, one-half the Company facing the wagon road and the other half at right angles facing the railroad.

The enemy demonstrated in force on our front, facing the wagon road, but attacked vigorously on our other front, paralleling the railroad, three times.

Before the third attack, when about two hundred yards from our line, the Federal officers halted their line of battle and placed themselves in front of their men, sword in hand, and seemed to be haranguing their troops and exhorting them to a supreme effort to carry the position held by the Confederates. When the advance was resumed, the officers remained in front, indicating a struggle *à l'outrance*.

Captain Humphrey, who, all through the battle had been restlessly walking to and fro, seemed now to become quite composed, remarking cheerily: "Now, boys, you will have a chance to show your training. I want you to aim straight and fire fast; but wait for the command to commence firing."

This caution prevented any premature firing, and the enemy advanced to within seventy-five yards of us, when their officers began to retire through the ranks and take position in the rear. Seeing this, private Ben Schipman, of the Cadets, yelled out, "Stay in front if you want to get those Yankees here!"

This pleasantry evoked the rebel yell all along the line of the Sixth, to the music of which the firing commenced, and the last attack of the Federals was repulsed.

After the battle, the enemy's dead were thicker in front of that part of the line held by the Cadets than anywhere else.

Captain Humphrey was wounded in the leg, Lieut. Aldrich in the right thigh, left arm, and right shoulder; Sergt. Simms had a finger of his left hand shot off; Corporal Hodges was shot in the breast, and privates Gladney, Quattlebaum, and Hodge were wounded in the right leg, right side, and left arm respectively. There were other casualties in the recruited portion of the company.

The Battle of Trevilians was fought on Sunday, June 12th. On the following Wednesday, the Cadets took part in the fight at White House, and ten days later in the action at Riddle's Shop.

The Sixth Regiment then had a breathing spell of a month, at the end

THE STORY OF THE CITADEL 69

of which time it bore the brunt of the sharp fight at Lee's Mill, in which Corporal Dozier was wounded and private Schipman captured.

The next fight in which the Cadets figured was Gravel Run, on August 23. This was followed by the hard-fought battle of Burgess' Mill, in which Lieut. W. A. Nettles was killed by a round shot, the same killing private Summers.

Lieut. Nettles was a gallant officer and splendid soldier, possessing the affection and respect of his fellow Cadets and the confidence of his commanding officers. The chronicler of this brief history of "The Cadet Company" must pause long enough in his story to say that he has never known in youth or manhood a nobler spirit than this boy soldier, who gave his services gladly to the State whose military school had fitted him to render efficient aid in her time of dire need, and finally his young life, with the resignation that characterized those sons who had graduated in the common Alma Mater when called to make the supreme sacrifice.

Speaking of graduates and non-graduates of the Citadel, it has been said that every graduate of the institution has been accounted for in some capacity in the Confederate service. This chronicle has been written to show that even those who were not alumni, and straggled out of ranks, yet *straggled to the front* in the "Lost Cause," and acquitted themselves as true sons of the Mother.

The battle of Burgess' Mill was the last fight in which "The Cadet Company" took part in Virginia.

Its next fighting began at Congaree Creek, in South Carolina, in February, 1865, where its division commander, Gen. M. C. Butler, decided to meet Gen. Sherman's Army in its march through the State. From Congaree Creek to Bentonville, "The Cadet Company" skirmished and fought incessantly. Butler's Division formed the rear guard between the points named.

On the tenth of March, 1865, our command surprised Gen. Kilpatrick's camp about daybreak, and the battle which followed lasted the whole day —no infantry on the Confederate side being employed. It may not be out of place to relate here one of the many episodes that befell the Cadets during their service, and the writer will record an incident of this battle in which Cadet "Shaftsbury" Moses measured sabres and fists with one of Kilpatrick's troopers. The Cadet Company was fighting hand to hand with the enemy, and Moses' horse was killed under him. On freeing himself from his dying horse, he found himself confronted by a big Yankee, sabre in hand.

Moses, being a smaller man than his adversary but dead game, determined to force the fighting, and made a furious rush inside his adversary's guard, which caused a clinch and a fall—"the Gael above, Fitz James below." Not only so—the Gael had, in the brief struggle, got a firm hold with his teeth on Fitz James's (Moses') finger. As good luck would have it, Private Bill Martin, whose horse had also been killed, came along in the nick of time and in his own expressive language, "lifted the Yank off of Shaftsbury" with his revolver.

As no such name as "Shaftsbury" Moses appears on the muster roll of the Cadet Company, it is proper to state that Cadet J. H. Moses while at the Citadel, on account of his scholarly style of composition had been dubbed by his fellow cadets "Lord Shaftsbury," and he carried that nickname through the war, many of his associates not knowing that it was not his real name.

In this battle, Sergt. G. M. Hodges' horse was killed under him, and he was shot in the side. Though wounded, he succeeded in capturing another horse and continued in the battle until disabled by a wound in the shoulder. Strange to relate, after the battle investigation showed that the enemy's bullet had entered the same hole in his coat that was made by the one which wounded him at Trevilians on June 12, 1864.

In this battle with Kilpatrick's troops, Captain Humphrey also was wounded in the arm by a grape-shot while charging a battery. He was carried to the hospital at Raleigh where the surgeons informed him that his arm must be amputated. He refused to submit to the operation from a morbid horror of going through life maimed, and died a short time before Lee's surrender.

Cadet Humphrey illustrated the noble worth of the Citadel as an educator of the youth of the State as signally, perhaps, as any case in the annals of the institution. He entered its walls poorly prepared by a brief period of schooling, but by dint of earnest application and a lofty ambition to succeed, from a position near the foot of the class the year of his matriculation, he stood at the head of his class the year he left.

Humphrey was gifted with a fine intellect and every natural quality to make for himself a successful career. Fate willed otherwise than that he should survive to fulfill the promise of his youth, and after a term of service brief but brilliant enough to satisfy the dream of any Paladin of romance, he died just in time not to know that the good fight had been made in vain.

The Cadet Company fought in the battle of Bentonville, and learning that Johnston's army was to be surrendered, by permission marched out of camp the night preceding that event with the idea of making its way to the trans-Mississippi part of the Confederacy, but disbanded finally under counsel of its Colonel when he bade them goodbye. (End of account.)

The story of "The Cadet Company" preserves the record of a few of the ex-cadets who splendidly served their State. But the larger number of boys who came for a time to the Citadel and left with uncompleted records disappeared into the outside world and have been lost to sight and knowledge. And yet there were undoubtedly many of these forgotten cadets who nobly met the call of their country when they heard it. Countries now erect monuments to their "Unknown Soldiers." In some such spirit, we may pause to pay a tribute to one of our lost ex-cadets, strangely brought up out of oblivion to receive a meed of praise in these latter days. Let us call it

THE STORY OF A FORGOTTEN HERO

Many years ago (to be exact, on January 1, 1856), an up-country boy —from Newberry County—entered the Third, or Sophomore Class at the Citadel, after having completed a year at the Arsenal.

He was not an industrious student, and probably he got into much mischief, because at the end of the first year he stood not far from the foot of his class, and was the very last in "conduct." In his junior year, his record showed no improvement in studies, and he still had the unenviable distinction of having more demerits than any other member of the class. Indeed, he was permitted to rise to the senior class only "under censure" by the Board of Visitors. However, he did not take advantage of the Board's consideration, for he did not return to graduate, and his name, therefore, does not appear on the roll of the alumni. Like hundreds of others who sojourned for a space inside the walls of the Citadel, and whose names have sunk into oblivion, only this brief and inglorious record can be found in the annals of the institution where he spent three years of young manhood:

"Allowed to rise to the First (Senior) Class, but under censure for neglect of study and for conduct, December 1, 1857.

No further record of any kind.

The casual delver into history might pause after reading this truncated biography to ponder for a brief minute on the personality of this forgotten boy. Where did he go? What did he do? Was his life so fruitless that he deserved the oblivion into which his name had fallen?

By a curious circumstance, in the year 1926, an inquiry was instituted by the Gainesville (Ga.) *Eagle,* in connection with its 66th anniversary, as to the founder of that paper. And behold, our long-lost hero is called up out of the forgotten past. Here is an abstract of the account:

"When you think of the beginning of the Gainesville *Eagle,*" the writer rhetorically says, "your mind reverts to William Henry Jamison Mitchell, the young man who came here sixty-six years ago with the sunrise of Hope shining in his face and launched his barque on the uncertain sea of journalism.

"He was born at Spartanburg, S. C., April 5, 1836. When he was sixteen years old, he went to Newberry, S. C., and engaged himself to a printer in the capacity of devil and general utility.

"In 1855, he entered Citadel Military College at Charleston. In the junior class of 1857 his name appears on the roll with nineteen others. For some reason not now known, he and four other members failed to graduate. Six members of this class of 1859 were killed in battle in the Civil War.

"Young Mitchell came to Gainesville in the early spring of 1860, and began to write a history of the Cherokee Indians. Afterwards, when he decided to start a newspaper, he announced that this history would be printed in its columns when its circulation had reached 1,500.

"He borrowed $600 from Mr. J. H. Banks, his note being endorsed by Mr. E. N. Gower, who made the first wagon ever turned out in Gainesville. With this money he bought the proverbial shirttailful of type, a Washington hand-press, and a limited supply of the other necessary paraphernalia.

"The first issue was printed August 10, 1860.

"At that time the town had about 150 inhabitants, and everybody knew everybody. If a householder erected a new gate in place of the old one that had got tired and fallen down, it was a matter of town-wide interest. The women wore hoops and whalebone stays and cherished modesty as a precious jewel. The chief dissipations of the young people were the ice-cream sociables given by the foreign missionary society. There they played clapout and steal-partners.

"Young Mitchell responded admirably to the wooings of the good things of life, especially those in the fluidic state, which were then of an incomparable quality, and going at 75 cents per. An ancient citizen once told us that when he got illuminated, one could see to read by him for quite a distance. On these occasions, the youthful Lyman Redwine, who was employed as devil to roll the forms, make fires, and chaperon the office towel, in addition to these onerous duties then performed as general manager.

"By the fall of 1860, the dark clouds of war were already lowering over the land at the prospect of the election of Lincoln.

"In the issue of October 26, 1860, the editor mentions the formation of a military company in an adjacent county, and urged preparedness on the part of Hall County people. He said, 'Wisdom teaches us to prepare for war in the time of peace. The time may come when our people will see the folly of neglecting this advice.'

"The writer's presentiment proved only too true.

"In the early spring of 1861, the dread alarms of war were sounded, and the young editor heard the call of his country. He raised a company and was elected captain. It became Company A, Eleventh Georgia Regiment. For several months they drilled on a vacant lot in the town. By command of the captain, every member of the company learned to cook.

"While in training, a beautiful silken banner was presented to the Company by Mrs. Richard Banks. It was made in Charleston by ladies imbued with Southern sentiment.

"And there is a romance connected with Capt. Mitchell and his short residence here. A romance in which War claimed its tribute of a broken heart. While the men were being drilled, Captain Mitchell visited the country home of one of his men, where he met Miss Elizabeth Drusilla Hudson. They fell in love, and were engaged to be married when the war should end, if the Fates would spare the life of the gallant Captain. On the day the company left, Miss Hudson was in town to bid her lover farewell. As he marched with martial step along the thoroughfare, he looked back and saw her face as she waved her hand. And that was the end of it.

"The company left here early in July 1861. The gray clad patriots who marched away so cheerfully on that bright July day, sixty-five years ago have all gone to a far country to hold reunions on fairer fields—all except one, Corporal Robert Clinton Young, who has thus far been spared by the fierce spirit of the scythe and hour-glass. He is now living at the home of Mr. John Simmons, and is cheerful and chipper at 91.

"When we started out to write this story, with the desire to tell as much as possible about Captain Mitchell, we wanted to find out something about his personal appearance. But the old 'uns are about all departed. Of the ten thousand present inhabitants of Gainesville, not one remembers anything of the founder of the *Eagle,* except Corporal Young. He said Mitchell was a small man, weight about 130; light complexion and fair; not at all handsome, but a good man and loved by his company.

"The company rendered splendid service in Virginia and suffered many casualties at Malvern Hill, July 1, 1862, Second Manassas, August 30, 1862, and at Sharpsburg, September 17, 1862. At the Second Battle of Manassas, Captain Mitchell was slightly wounded, and his First Lieutenant, James C. Gower mortally wounded, dying at Warrenton, Va., on September 22nd. A vivid description of the battle, and the death-wound of his lieutenant and friend is given by Captain Mitchell in a letter to the mother of Lieutenant Gower, written in camp on the Rapidan, November 16, 1862.

"And then at Funkstown, Maryland, July 5, 1863, while the army was on the retreat from Gettysburg, Captain Mitchell, at the age of twenty-seven, met his death in battle.

THE STORY OF THE CITADEL

"Captain William Henry Jamison Mitchell
April 5, 1836
July 5, 1863
Died For His Country"

✦ ✦ ✦ ✦ ✦

The Battle of Tulifinny

The only engagement in which the Battalion of State Cadets formally took part was at Tulifinny Creek, on the Charleston and Savannah Railroad, about six miles below Yemassee Station, in December, 1864.

The coastal country of South Carolina is deeply indented with saltwater sounds, tidal rivers, and wide swamps, and presents a difficult terrain for offensive military operations.

The Federals had early in the war captured Port Royal, but no very serious effort was made to advance up the coast until Sherman reached Savannah after his march to the sea, and it was not until the latter part of the year 1864 that the Federal forces began their systematic advance to cut the line of the Charleston and Savannah Railroad towards Charleston.

Above Parris Island, the wide extension of Port Royal Sound is known as Broad River. About ten miles north of Beaufort, it comes to a head in three tidal rivers winding through wide salt marshes which spread like three great distended fingers six or eight miles further inland, where they finally degenerate into shallow black streams lost in impassable swamps.

These arms are the Coosawhatchie, Tulifinny, and Pocotaligo rivers.

The railroad crosses these rivers about the head of navigation for small boats, and it was the railroad which was the objective of the Federal troops in the Tulifinny engagement.

Of the engagement at Tulifinny, we shall quote the official report of Major White, who was in command of the battalion.

Report of Major J. B. White

On the 30th November (while the Citadel Cadets were in camp at Orangeburg, S. C.), I received an order from the Adjutant and Inspector General of South Carolina, through the Chairman, directing that I "proceed and report to the nearest Confederate General for service in the field." At Branchville, orders were received from Governor Bonham changing the destination of the battalion to Charleston, on reaching which place I reported to Major General Samuel Jones commanding that Department.

Captain J. P. Thomas here reported to me with the Arsenal Corps. The command was then fully organized, the Citadel Cadets forming Co. "A," and the Arsenal Cadets Co. "B." All the non-commissioned officers of Co. "B" were furnished from the higher (Citadel) classes. The officers of the Battalion were Lieut. Amory Coffin, Adjutant; Lieut. A. H. Mazyck, Quartermaster; Captain H. S. Thompson and Lieutenants N. W. Armstrong and J. E. Black, assigned to Company "A"; Captain J. P. Thomas and Lieutenants A. J. Norris and R. O. Sams to Company "B."

On the evening of the 3rd of December, the Battalion was ordered to Coosawhatchie, on the Charleston and Savannah Railroad, but was stopped at Pocotaligo.

December 6th, the enemy threatened the railroad by way of Gregory's Point, and the Battalion was posted to protect Tulifinny trestle, but was afterwards ordered to attack the enemy, who was then engaged with our forces in front, on his right flank. These falling back, the order was countermanded and the command resumed its position on the railroad.

December 7th, I was directed by Col. Edwards, of the 47th Regiment, Georgia Volunteers, to take Company "A" of the Battalion, with other troops, and advance upon the enemy in order to ascertain his exact position, and determine the propriety of attacking him with the forces at hand.

The entire line of skirmishers soon became engaged with those of the enemy, but advanced steadily, driving them back upon their entrenchments, in front of which the line was halted, and after accomplishing all that was desired, fell back in perfect order.

During this skirmish, which lasted about three hours, Company "B" relieved Company "A" (whose ammunition had become exhausted), so that the entire battalion became engaged.

This was the first time the Battalion of Cadets met the enemy, but their conduct was such as to excite the commendation of the veteran troops by whose side they fought, and to call forth the approval of the commanding general as well as the colonel commanding the expedition.

Every Cadet acted with conspicuous gallantry, and showed that the Academy had made him a thorough soldier for the battlefield.

December 9th, the enemy advanced against our position, but as the attack was directed principally upon our right, and the Battalion being on the left, it became only partially engaged.

The casualties on Dec. 7th were as follows: Lieut. Amory Coffin, Adjutant, wounded in the head (severe); Cadet J. B. Patterson, mortally wounded; Cadets J. W. Barnwell and E. C. McCarty, severely wounded; and Cadets S. F. Hollingsworth, A. J. Green, A. R. Heyward, and W. A. Pringle, wounded slightly. All these were in Co. "A."

The Battalion went into camp at Tulifinny until Dec. 25th, when it was ordered to James Island.

While in Camp Tulifinny, Cadet Palmer, of Co. "B" had his left hand shot off by a shell from the enemy's battery.

I regret to report, also, the deaths of the following Cadets, all of whom died from diseases induced by the exposure and hardship of service, viz., R. F. Nichols and John Culbreath, of Co. "A," and G. O. Buck, T. A. Johnson, and R. Noble, of Co. "B."

I would take this opportunity to express my obligations to the officers under my command for the zeal, ability, and alacrity with which they discharged their duties. Nor can I fail to call to your attention those young but noble sons of our State. Upon the battlefield, in camp, on the march, on picket, or working upon defences, they were ready for every emergency; manifesting at all times, and under the most trying circumstances, a manly and soldierly aspect, not finding fault with those in authority, but doing their duty cheerfully and well.

And now let us supplement Major White's official report with an account written sixty-five years afterwards by one of the younger Cadets in the battalion at Tulifinny, whose memory of his experiences has not been diminished by the years.

My Recollection of Fight at Tulifinny Creek, South Carolina, in December, 1864.

By *George M. Coffin*

In January, 1864, our class of cadets, boys between 15 and 18, assembled at the Arsenal barracks at Columbia, S. C. to take up the studies and military training as fourth classmen of the South Carolina Military Academy, expecting to pass the year there before going on to the Citadel at Charleston to continue the four-year course.

But the War Between the States was going on, and gradually drawing to a close, and by October or November our corps was being used to guard the camp of Union prisoners at Columbia and the State House grounds. Late in November or early in December, 1864, when the South in its extremity was "robbing the cradle and the grave" to put soldiers on the fighting line, the Arsenal cadets were sent down to Charleston and combined with the Citadel cadets to form a battalion under command of Major "Benny" White, 343 strong. The writer had been third sergeant in his class, but when he joined the Citadel battalion, those chevrons had to come off when he took his place as a private in the ranks. Very soon the battalion was sent in box cars to Pocotaligo where we were just going into camp and to get something to eat, when we were suddenly ordered to march further down the track of the railroad between Charleston and Savannah which Union troops were trying to cut. After marching and countermarching all afternoon and evening, we finally halted alongside the railroad near Tulifinny Creek. We had come without overcoats or blankets, and it was raining, and without breaking ranks we lay down company front on the cornbeds and tried to sleep. I remember putting my plate canteen over my face to keep the rain off. I think we had some bacon and "hard tack" to eat, after coming away without any midday meal.

The next morning, Dec. 7, at daybreak we marched across the railroad track into the woods where the fight with the Union troops took place. The Citadel company went in first and bore the brunt of the action.

The Arsenal Company was held in reserve. We lay down to avoid the bullets and shells passing over us, and in a heavy rain on some swampy ground. We were, after an hour or so, sent in on the firing line to relieve the Citadel company which had suffered considerable casualties. One cadet, Patterson, fatally wounded, and Lieutenant Amory Coffin with cadets Barnwell, Green, Heyward and some others wounded. Our company stayed in a while firing at the Yanks, though I did not see any. We had muzzle-loading Springfield rifles, and after each shot had to tear cartridge, ram down bullet, restore ramrod and put on percussion cap before we could fire again. The enemy must have been driven back by the Citadel Company, because after a while we were withdrawn to some rifle pits in an open field. There we stayed until night, when we were relieved and went into camp with fires to warm us and dry our wet clothes. And my! it was good to get something warm to eat again, for I do not remember having had anything but some "hard tack" and bacon the night before. After that we went back to a camp alongside the railroad on the other side, in some wood adjoining a "broom grass" field. We had not at any time any tents or other shelter, and lived in the open air. We did our own cooking, and the first biscuits I made without any "rising" laid me out under the trees for several days. Early in December the Arsenal class of 1865, boys of 15 to 18, were brought down to our camp, and the elder cadets were used as drillmasters for them. The Federals were getting the range of our camp, and one day while the drill ground was covered with squads, shells began to tear up the ground. Fortunately no one was hurt. About the same time a shell passing right through our camp took off a hand of "Big Bill" Palmer of our Arsenal class. After the Tulifinny action we did picket duty. I remember standing alone for hours in an open field on picket watching for enemy sharpshooters in the tall pines and thinking that I was only 17 and did not want to die yet. The only other action we had after Tulifinny was one morning early we were hurried to some breastworks on the other side of the railroad track and fired several volleys at an enemy we did not see. Later in the day picked cadets stood on the breastworks and did some sharp-shooting. We heard that one dead Federal with red beard had been found in the woods in front of us.

A little later as the enemy was getting range of our camp and drill grounds and we had no tents or shelter, the Arsenal class of 1865 was sent back to the barracks at Columbia to go on with their studies and be drilled. They were under command of Captain Thomas, Lieutenants Patrick, Norris and Sams, and Cadets Murdoch, Lamar and I went along as drill-masters. We spent one night at the Citadel in Charleston, and in our room I found an excellent rope-ladder very suggestive of its use for breaking garrison limits. (End of account.)

THE FALL OF CHARLESTON

After four years of gallant defense, the end came without any of the elements of the pomp and panoply of war. It was drab rather than dramatic.

On February 18, 1865, the Federal troops, under the command of Lieutenant Colonel A. G. Bennett, marched unopposed into the city and hoisted the United States flag over the customhouse, the Citadel, and the Arsenal.

There was no guard at the Citadel to surrender to the conquerors—only the venerable professor of science, Dr. William Hume, who was left by Major White in charge of the buildings when the corps of cadets went into the field in the preceding December. The conquerors kept the old gentleman up until 2 a.m. as a hostage, in case the building might be blown up by Rebel treachery.

Dr. Hume, the one member of the ante-bellum faculty of the Citadel, with a service practically coterminous with the operation of the Academy—having been elected a professor on May 20, 1844, and being the lone officer in charge to surrender it to the conqueror—was a remarkable character. It will be of interest to the reader to have inserted here the following sketch of him contained in Colonel Thomas's history:

Dr. William Hume was born in Charleston, S. C., July 26, 1801.

His father was John Hume, a rice planter, who served with General Francis Marion in the Revolution. His grandfather, Peter, son of Robert Hume, of Berwickshire, Scotland—of the family of the great historian—was born in England, and came to America about 1725. He married a lady of Londontown, Maryland, and removed to South Carolina.

His mother was a daughter of William Mazyck, grandson of Isaac Mazyck, a Huguenot refugee who came to America in 1686.

Dr. Hume's early education was conducted chiefly at home under tutors; but he was graduated at Yale, and took his medical degree in New York. He then went to Europe, where he spent some time in travel and study. He was a student at the Garden of Plants, Paris, under Valenciennes and

Geoffroi St. Hilaire, who were co-workers with Baron Cuvier, and there met Humboldt and others of like distinction. He was engaged for a time as dresser at Guy's Hospital, London, under Sir Ashley Cooper.

While in England, Dr. Hume married Catherine, daughter of Jonathan Lucas, of Deptford. His second wife was a sister of the first.

Returning to South Carolina, Dr. Hume practiced his profession in Charleston, and planted on lower Santee. He engaged in some business ventures for which he was little suited. Money-making was foreign to his tastes, which were essentially literary and scientific. The truth in nature he loved and earnestly sought, and loved and sought it for its own sake.

In 1844, he was elected "third professor" at the Citadel with the rank of second lieutenant, Major Colcock being superintendent and Lieutenant F. W. Capers "second professor."

He was soon relieved of all military duty and assigned to the department of chemistry, geology, and mineralogy—subsequently called the department of experimental science, comprising chemistry and physics.

Professor Hume was now in his element and out of trouble, since he was relieved from the military which be abhorred and for which he was in all respects unsuited. To command on drill or dress parade had been to Lieutenant Hume not unlike commanding an unruly elephant. His military *faux pas*—such as being run over by his platoon on drill, and his vocabulary on dress parade, such as, "Gentlemen, will you be kind enough to *present arms*"—these among the traditions of the cadets he taught. In a word, the truth of history compels the statement that as a military man, Lieutenant Hume was an egregious failure. His office fitted him no better than his uniform—and that was always peculiarly a misfit.

But it was far otherwise with Professor Hume in his laboratory and in the chair as lecturer and teacher. Here he was a shining light. What he lacked in military attainments he made up in scientific accomplishments. Nor did he confine his labors and scientific spirit to the walls of the Citadel. He looked to the interests of Charleston and the State in the spirit of enlightened and patriotic citizenship. The question of the public health of the city engaged his attention, and his contributions to the subjects of sewerage and epidemics were useful and creditable to his genius. He sought to make chemistry and natural science minister to the public good. He served for several terms as a member of the city council, and was a useful man in his day, as well as a sincere philanthropist.

As a professor at the Citadel, he gave scientific prestige to the school. He was a most pleasing lecturer, having at his command an ample supply of good, pure English. He was an enthusiast in his department, which he invested with much interest to his cadet classes.

Dr. William Hume had many virtues, and no cadet who was under the spell of his charming lectures will fail to think kindly of his former quaint preceptor.

⚜ ⚜ ⚜ ⚜ ⚜

The scientific instruments of the departments and the library books had previously been sent to Columbia for safe keeping under the mistaken impression that Sherman, in setting out from Savannah after his famous march to the sea would go north by way of Charleston, in order to wreak vengeance on "The Cradle of Secession."

Perhaps the Charleston people had got wind of the letter of General Halleck, chief of staff at Washington, addressed to General Sherman at Savannah, on December 18, 1864, suggesting that "should you capture Charleston I hope that by some accident the place may be destroyed; and if a little salt should be sown upon the site, it may prevent the growth of future crops of nullification and secession," to which pleasantry General Sherman had replied in a similar vein: "I will bear in mind your hint as to Charleston, and I do not think 'salt' will be necessary. When I move, the Fifteenth Corps will be on the right of the right wing, and their position will bring them into Charleston first; and if you have watched the history of this corps, you will have remarked that it generally does its work pretty well."

Fate must have smiled ironically when Sherman's hordes left Charleston unharmed but sacked the capital city, and destroyed or carried off all the valuables which had been sent there from Charleston for safe keeping. Somewhere, perhaps, an astronomical telescope, which had formerly been mounted in a dome on the southwest bastion tower of the old Citadel, may still be searching the skies; and other instruments survive as relics of the war on the what-nots of old G.A.R. veterans; but, with one exception, all the property of the Citadel sent to Columbia was either destroyed or lost. The exception is a military relic, a fifteen-inch shell which had been fired by the Federal fleet at Fort Sumter in April, 1863, and which had been presented to the Citadel by General Beauregard at the commencement exercises that year. The story of this relic is told by the inscriptions on the pedestal on which the shell rested for fifty years in the museum at West Point.

  1. Fifteen-inch hollow Shot
   fired by the Abolition Fleet
   of Iron clads at Fort Sumter.
     April 7th, 1863
  2. Presented to Citadel Academy by
   General G. T. Beauregard.
    Charleston, S. C. April 27th, 1863

3. Taken at Columbia, S. C. Feb. 17th 1865 by the Troops of the U. S. under Major General W. T. Sherman.
4. Presented to the U. S. Military Academy by Major General Wm. B. Hazen.

April 1, 1865

5. RETURNED TO THE CITADEL
by order of
HON. HENRY L. STIMSON
SEC. OF WAR,
April 4, 1913

In 1913, upon the request of the superintendent of the Citadel, Honorable Henry L. Stimson, Secretary of War, returned this trophy to the Citadel, and it now rests upon its pedestal in Memorial Hall.

Lieutenant Colonel Bennett had evidently not received special instructions from General Sherman; for his first care was to insure good order in the city. His second general order, issued from the Citadel, read as follows:

"HEADQUARTERS U. S. FORCES
Charleston City, S. C., Citadel, Feb. 18, 1865.
General Order No. 2.

"Hereafter, the sale of all malt or alcoholic liquors is strictly prohibited.[1]
By order of Lieut.-Colonel
A. G. BENNETT."

A few days later, Colonel Bennett changed his headquarters to the former residence of Judge Mitchell King, at the northwest corner of Meeting and George Streets; but a detachment of Federal troops was left at the Citadel, and remained in possession of it for a period of seventeen years.

The Arsenal Academy, in Columbia, fared even worse.

The last class to matriculate in the South Carolina Military Academy was received by Captain J. P. Thomas, at the Arsenal Academy, on December 14, 1864.

These 114 boys ranged in age from fifteen to eighteen years. A few of them had already seen service in the Confederate Army, and one of them, A. F. O'Brien, of Colleton, had lost his right arm in battle. Many years afterwards, in 1889, Mr. O'Brien was one of the three Confederate soldiers selected to represent South Carolina at the funeral of Jefferson Davis.

---

[1] A necessary regulation. The *Courier*, of February 22, states: "A soldier belonging to the 54th Massachusetts C. T. was shot last evening in front of the Citadel, and, it is believed, mortally wounded. He was under the influence of liquor, and had been unruly, refusing to obey the officer of the guard."

The new class began their studies, but academic work shared time with military duties. The Arsenal Cadets did provost guard duty in the town—not an easy task when such unruly soldiers as Wheeler's men were around—and guarded the seven hundred Federal officers who were confined as prisoners of war in the grounds of the Insane Asylum.

On February 16, when the advance of Sherman's army made the evacuation of Columbia imminent, they were guarding the highway bridge across the Congaree River at the foot of Gervais Street, and were under fire of the Federal sharpshooters on the Lexington side, but withdrew only after setting fire to the bridge and destroying it.

"Captain Thomas, the Superintendent, received orders late in the evening to evacuate the Arsenal and join Gen. Garlington's command of State Troopers, which was on the eve of retreating from the city. At the last moment the idea came into his mind that, of all the property under his charge, he could at least save the Flag of the Citadel Corps. Accordingly, he ordered it brought out, and placed it in the keeping of the Battalion of Arsenal Cadets."

The next day, the Arsenal Cadets marched out of Columbia, attached to the wagon train of General Garlington's Brigade as rear guard, and when Sherman's Army burned Columbia the barracks and academic buildings of the Academy were destroyed in the general conflagration.

Thus was brought to a close, on February 18, 1865, the operations of the two schools composing the South Carolina Military Academy.

This date marks the end of the first period of its history.

The two homeless battalions of boys, however, retreating into North Carolina ahead of Sherman's Army, continued to operate as military units in the field for more than two months longer.

About March 1, 1865, the Citadel Cadets performed their last military service for the tottering Confederacy by escorting a body of Federal prisoners to Raleigh, North Carolina, and were then recalled to South Carolina by Governor Magrath, whose headquarters were at Spartanburg.

Meanwhile, the Arsenal Cadets, with General Garlington's militia brigade in their northward retreat from Columbia, passed through Winnsboro and White Oak, and then crossed the Wateree River, expecting to get out of the way of Sherman's Army. At Lancaster, however, General Garlington found himself pressed by the advance guard of the Federal cavalry. It was a case of *sauve qui peut*. He turned over the Arsenal Cadets to their superintendent with instructions to report to Governor Magrath. Easier said than done. But Captain Thomas put his cadets in light marching order, crossed Lynch's Creek, and in one day was at Ansonville, North Carolina, where his tired and hungry boys were bountifully fed by the hospitable ladies of that village. Turning westward, Captain Thomas finally got the battalion out of the track of the Federal Army, and rested at Concord, from which point the foot-weary young soldiers were glad to get railroad transportation to Charlotte and Chester, South Carolina. From the latter place, there was a cross-country march to Spartanburg, where the Citadel Battalion was already encamped.

The Arsenal boys arrived in Spartanburg on March 8, and were quartered in Wofford College buildings for two days. It was here that the two battalions of cadets met again, having last been together at Tulifinny, and Captain Thomas took the occasion to return to the Citadel cadets their flag which had been sent to the Arsenal for safe keeping in the preceding February.

On the 10th of March the Arsenal Cadets were furloughed for fifteen days, scattering to their homes in various parts of the State by such means of transportation as they could procure. The marvelous thing was that at the end of the fifteen-day furlough they all, practically, reported back for duty.

During the last week of this month, the two battalions left Spartanburg and marched to Greenville, where Governor Magrath had moved his headquarters, and they were in camp there when the news arrived of Lee's surrender.

A curious incident in connection with the historic field of Appomattox is related in Colonel Thomas's History:

"In the Washington correspondence of the *News and Courier* of date of January 24, 1883, there appears the following:

'I have recently come into posession of a relic that I prize very highly. It is but a small square piece of toweling, but it is a section of the white flag of truce which floated between the lines of war at Appomattox, the signal at the sight of which two great armies rested upon their arms, never again to renew the greatest struggle the world ever witnessed between men of the same race and nativity.'

"The story of this flag of truce is of special interest since it links a Citadel graduate with the virtual end of the War Between the States. It was Major Robert M. Sims, of the Class of 1856, then on General Longstreet's staff, who bore that famous flag of truce, he having been sent by General Longstreet to say to General Gordon that if the latter thought proper he might send a flag of truce to General Sheridan for a suspension of hostilities until General Lee, who had gone to meet General Grant, could be heard from.

"Major Sims, upon delivering his message, was requested by General Gordon to take the flag. This Major Sims did, exposing himself unshrinkingly to a shower of bullets. But the firing soon ceased, and Lieutenant Colonel Whittaker, of the 1st Connecticut Cavalry, conducted Major Sims to General Custer. His work being done, Major Sims, retaining the towel used as a flag of truce, returned to General Gordon who requested him to bear it to another part of the line. Major Sims excused himself, as he conceived it to be his duty to report back to General Longstreet. General Gordon thereupon sent Major Brown, of his staff, who borrowed the historic towel of its owner. Major Sims never saw it afterwards, but was told by Major Brown that he had loaned it to one of the Federal officers to exhibit in order to prevent being fired upon as he entered his own lines. The towel seems to have fallen into Colonel Whittaker's hands since he, as *General* Whittaker, nearly twenty years after the incidents narrated, gave it to the Washington correspondent of the *News and Courier*."

Two weeks after Appomattox, the army of General Johnston, the last body of troops capable of offering any serious resistance to the Federal forces, surrendered at Bentonville, and the military conquest of the South was complete.

It was on the next day that the Board of Visitors of the Citadel met in Greenville. Remarkable to relate, the Board—far from accepting the situation as hopeless for the two Academies—made plans for increasing the number of cadets in both battalions. It was immaterial that the Citadel was in the hands of Federal troops, and that the Arsenal had been destroyed by fire; "the headquarters of both these institutions would be established at such places as the Chairman, in consultation with the Governor, might select!"

The Citadel Cadets, who had been in the field since the previous December, were on April 29 granted a fifteen-day furlough. The Arsenal Cadets were to be kept for a time in their log huts at Greenville and recruited. However, on May 1, being advised of the approach of Stoneman's raiders, Captain Thomas broke camp and set out with his two companies of cadets, 150 strong, for Newberry.

A few miles out from Greenville, the Arsenal battalion became again the custodians of the Citadel flag, Lieutenant Lanneau, with whom it had been left in Greenville, when the Citadel Cadets were furloughed, riding out and delivering it to Captain Thomas.

And now, one of those queer tricks of Fate happened.

We have seen that on January 9, 1861—three months before the War began—it was a detachment of Citadel boys who fired the first hostile shot at the American flag when the steamer *Star of the West* attempted to relieve Fort Sumter. And now after four years of bloody strife, when the armies of the Confederacy were all dissolved and the war was over, the little band of Arsenal Cadets—the freshman class of the South Carolina Military Academy—was to fire the last hostile shot.

On this first of May, after a weary march, not far from Williamston, near the spot where the Piedmont Cotton Mills now stand, the boys were resting—and some were asleep—when the bivouac was startled by a sudden dash of the cavalry company composing the advance guard of Stoneman's brigade. The spot where the Cadets were resting was around a curve in the road, and it is likely that the raiders were as much surprised as the Cadets when they rode thus unexpectedly upon a body of young men in uniform. They fired only a few shots and dashed on down the road, the affair lasting only a few minutes. Captain Thomas says: "Some confusion incident to a surprise ensued; but when the flag was displayed in the road, the command promptly rallied around it, and the fire was returned by the Cadets with effect."

Writing of the incident sixty years afterwards, Captain E. A. Smyth, who was one of the Arsenal Cadets says: "Only a few shots were exchanged. When the raiders retired, but one man fell off his horse and lay on the ground, and was picked up by a comrade and flung across the front of his saddle. We thought he was dead; but the whole affair lasted three or four minutes, though it seemed to us cadets a much longer time."

Although the episode may be considered nothing more than a shooting up of a boy camp by a band of irresponsible ruffians, the ruffians rode on with the loss of one man in a Federal uniform. And so, the boys who had fired first on the flag were in all probability the last to inflict a casualty upon the armed forces of the United States government in the War of Secession.

A week later, at Newberry, the governor ordered Captain Thomas to furlough his command. The order read, "for fifteen days"—but all knew when they said good-bye that this was the end.

Dolorous Days

In the history of the United States, following the War of Secession, the decade from 1866 to 1876 is usually referred to as the Era of Reconstruction. A recent historian[2] has more aptly termed it the Tragic Era.

Certainly, in South Carolina, the nouns "destruction," "desolation," and "destitution" better express the condition of the prostrate State than "reconstruction."

In the first year or two after the War, there was a natural uncertainty as to how adjustments were to be made in restoring the seceding states to their fellowship in the Union. The assassination of President Lincoln was

[2]*The Tragic Era,* by Claude G. Bowers (foretold by Calhoun in 1849).

generally recognized in the South as most unfortunate for this section, but historians are not at all sure that even Lincoln could have curbed the implacable element in his party represented by the ilk of Thaddeus Stevens, and it is quite conceivable that the exalted place in history occupied by the martyred President might never have been his had he lived through his second term.

While it may be assumed that a large element in the North was willing for the Southern states to resume their previous status in the Union upon the acceptance of the principle of the indissolubility of the Union and the abolition of slavery, the more vocal of the politicians were determined upon a policy of previous punishment and the adoption of some method of insuring the perpetuity of the Union (Republican) Party in power.

The latter was the determining consideration in the minds of the most influential element of that organization. If the Southern states resumed their former status as coequals in the republic, the abolitionists believed the negro would be kept in serfdom, and viewed with alarm the likelihood of a solid Democratic South holding the balance of power in the next presidential election. What was to be done?

The plan which was devised, worked out, and ultimately carried through was so diabolical in its results that Christian charity requires us to believe that not even the originators of the scheme foresaw its consequences. It was a deliberate policy of disfranchising the intelligent citizenship of the states "lately in rebellion," enfranchising the negroes, and establishing governments therein in which an inferior race, just out of bondage, and not far removed from African savagery, should be placed in power to rule a people who had won their independence from Great Britain and were noted for their culture and ability in statesmanship and government. Thus was the Republican party to be established permanently in control of the national government.

Under the protection of the military forces of the Federal government, the State was soon the prey of unprincipled carpetbaggers from the North and the more despised scalawags from the South, white men who were in the game for personal profit.

The orgy of corruption and crime, extravagance, robbery and debauchery, which ensued may not have been foreseen, but it assuredly was vile enough to satisfy the demands of the most implacable partisan hatred.

The worst feature of the period was the destruction of the friendly relationship which had existed between the former slaves and their masters, and the forming of an ever-widening breach of antipathy between the two races which has never been bridged.

A wise provision of the future would have dictated a policy of conciliation and sympathy. Perhaps this was too much to expect. At any rate, the Radical party laid a good foundation for the Solid South.

The General Assembly which met in Columbia in July, 1868, was composed of twenty-one white and ten colored senators, and forty-six white and seventy-eight colored representatives. This legislature ratified the Fourteenth Amendment.[8]

The constitutional convention of that year was, of course, under the control of the Radical element, and adopted a provision declaring that all schools and colleges supported by public funds should be "free and open to all the children and youth of the State without regard to race or color." The Act of 1869, reorganizing the University, contained a similar provision.

Some of the counselors of the colored people advised them against any attempt to mix the races in school or college. Governor Orr, in his message to the legislature in 1868—and on other occasions—urged the separation of the races, and suggested that the University be reserved for the whites and that the Citadel be reopened and converted into a college for colored students.

Fortunately for the Citadel, this recommendation was not carried out.

For a number of years, the negroes did not avail themselves of the opportunity offered by the Constitution to enter the University.

It was not until the year 1873 that the break came.

On October 7 of that year, a bright mulatto, named Hayne, applied and was admitted to the school of medicine of the University. Doctors LaBorde, Talley, and Gibbes, of the medical faculty, immediately resigned. The trustees, being determined upon a mixed policy, accepted their resignations and appointed more complacent professors. The negroes now entered in large numbers, and in the following year constituted about ninety per cent of the student body of two hundred.

---

[8]The General Assembly which met in Columbia in November, 1870, had a joint Republican majority of 118, and a joint black majority of 16.

CHAPTER VI
SIXTH DECADE: 1872-1882

IN Charleston, during these dark days, it was futile to think of recovering the Citadel and reopening it as an educational institution. In 1870 —five years after the War—a writer of a "Ramble About Town,"[1] thus describes the buildings: "The first building that strikes the eye after you cross Calhoun Street going up King Street is

THE CITADEL

"Here we see no upright cadets as in days of yore in their swallow-tailed gray coats, or their brown linen roundabouts, being drilled to march to the monotonous music of fife and drum, or strutting near the fence to stare at and be admired by the passing fair ones. No palmetto flag now asserts the dominion of South Carolina over the imitation fortress. The gay cadets are disbanded veterans, or occupy soldiers' graves, and the flag is folded among the mementoes of the past. But the Green looks even better than in the olden times, and the grim barracks, as ever, frown down upon it. The Citadel is a long, low, yellow rough-cast building, with two wings, each nearly as large as the main building. The interior of the western wing has been destroyed by fire since the War, but the injuries are not visible at a distance."

TO A VINE GROWING ON THE DESERTED CITADEL
By *Colonel John P. Thomas* (about 1875)

Twine around the little window
   Through whose bars I fain would gaze;
Let thy tendrils 'round it clinging
   Lift a voice in song of praise.

Cling yet closer in thy frailty
   'Gainst that aged, massive wall,
Keeping green and full of freshness
   All those memories I recall.

Telling me of hours departed,
   Full of joy and youthful glee,
Bringing back those happy faces
   That I nevermore shall see.

---

[1] Premium List, 1870, of the South Carolina Institute.

Once kind hearts, carefree and joyous
  Beat within those classic walls;
Many voices, long since silent,
  Answered to the tattoo calls.

There is silence now and darkness;
  Happy student days are o'er.
Alma Mater's halls are empty,
  Books are closed for evermore.

Still there linger recollections
  Of these dear ones tried and true,
And I love the small-barred window,
  Where the sweet vine ever grew.

At this time, the Federal garrison at the Citadel consisted of a battery of light artillery.[2] The junior second lieutenant of this battery was Charles William Burrows, who had just graduated from West Point. In 1920—just half a century afterwards—Mr. Burrows revisited Charleston, and gives this account of the Citadel of the early seventies:

"The officers of Light Battery "C," Third U. S. Artillery, to which I was ordered on graduating at the Military Academy in 1870 were as follows:

"Captain Wm. F. Sinclair,

"First Lieutenants Major Arthur (brother of President Chester A. Arthur), and John F. Mount,

"Second Lieutenants John M. Calif and Charles Wm. Burrows.

"The appearance of the Citadel as it is today is not very different on the outside from what it was in those earlier days. A fourth story has been added to the main building, which, however, affects its general appearance but little. As you examine the place and the grounds, however, a number of other changes are evident. A row of small buildings along the western side of the Square on King Street has disappeared. Also, a generation ago, a solidly built fence surrounded the entire Square. The Calhoun Monument was not there. The two present extensions on King and Meeting Streets did not exist at that time, and the old West Wing was roofless and in ruins, the cleared-up enclosure of which was used for stabling the horses of the Battery.

"Quarters for the married officers were at the front of the Citadel to the

---

[2] Subsequently—and possibly at this time—there was also at the Citadel an additional battery of foot artillery, serving as infantry.

right of the sally-port, while unmarried officers lived on the left side. The men were quartered in the East Wing.

"Yellow fever broke out in Charleston in August, 1871, and in consequence our Battery marched out to Summerville and camped there for about three months, breaking camp and returning to the city toward the end of November, when the early frosts had killed the fever."

The destruction of the West Wing by fire, to which reference is made, occurred on the morning of Saturday, October 30, 1869. It was not until twenty years later—October, 1889—that the smoked and scarred ruins had been cleared away and a new wing built and opened for faculty quarters.

In 1876, when the centenary of American independence was being celebrated by a great exposition in Philadelphia, the slowly reviving South was beginning once more to look up and hope.

A decade of Negro domination had exhausted the endurance of the white people of South Carolina, and a new Spirit of '76 was dawning in the hearts of its men. Red-shirted patriots began to organize all over the State, for a cause holier than any which had ever aroused them in the past—transcending in importance the issues of nullification and secession, because this meant the preservation of civilization itself.

But the redemption of the State was far from an easy task. It called for the ablest leadership that South Carolina could produce.

Fortunately, there were many men as wise as they were determined, and as patient as they were patriotic. During that most critical year in South Carolina history there were many incidents where it required all these qualities to save the State—such as the Hamburg riot, the Ellenton riots, and the Cainhoy massacre.[3]

In Charleston, where a number of negro Democratic clubs had been organized under the protection of the white people, there were serious clashes between these Hampton negroes and the Republican negro clubs. On one occasion, after a Democratic negro meeting had adjourned, the speakers, on leaving the hall, were surrounded by a yelling mob of black desperadoes belonging to the Hunkadory Club, and would have been severely handled but for the protection of fifteen white men who formed a hollow square "with the negro Democrats in their midst, struggled up King Street to the Citadel grounds, and delivered the objects of the mob's fury to the protection of the Federal soldiers there."[4]

As the political campaign drew towards its close, Governor Chamberlain

---
[3]In the Ellenton riots in September, 1876, two whites and fifteen negroes were killed, and eight whites and two negroes wounded. At Cainhoy, the negroes fell upon the unarmed whites, killing six and wounding sixteen others.
[4]*The Tragic Era,* by Claude G. Bowers, p. 511.

THE STORY OF THE CITADEL

requested additional troops to be sent to South Carolina, and on election day, November 7, it is estimated that 5,000 United States troops were distributed throughout the State.

When the election returns were canvassed, it was found that the State had given its electoral vote to the Republican candidates for President and Vice President, but that General Hampton had defeated Chamberlain for governor by a little over a thousand votes.

The radicals were desperate, and would not accept the result. In consequence, South Carolina presented the unusual spectacle of operating for several months under two rival governments, both Hampton and Chamberlain exercising the functions of chief magistrate, and separate legislatures enacting laws.

The former, however, had one very considerable advantage—as the Democrats were the taxpayers, the revenues were paid to the Hampton government.

Without money, the Republicans could not hope to hold on, even with the supporting presence of the Federal troops; and when the national government took these away on April 10, 1877, the carpetbag harpies took flight and the scalawags dropped into oblivion.

Thus, after a long night of hideous nightmare, the sun rose again.

A writer in the London *Times*,[5] in 1895, makes this comment on the close of the Tragic Era:

"In 1873, the conditions of life for the white population were becoming so utterly unendurable that the alternative to the civilized natives of the State was to regain possession of the executive and legislative government or to quit the country in a body. There was literally no other course, since men who are of Anglo-Saxon and Huguenot blood, inheriting the traditions of freemen, could not submit to live and suffer under a government scarcely differing from that of Hayti and San Domingo. The conflict was bitter but victory was won—by what means and at what cost we must not too closely inquire."

As soon as white supremacy had been reëstablished in the State, the alumni of the Citadel began to talk of reopening the institution.

In April, 1877, a small group of nine graduates met at the Charleston Hotel to discuss what measures seemed feasible for the recovery of the Citadel property, which had been in possession of the United States military forces for twelve years previous, and for the reopening of the school.

As a first step, it seemed desirable to reorganize the Association of Graduates, which had been *functus officio* since 1861, and this small first

---
[5]Thomas Edmonston, quoted in *History of South Carolina,* by Yates Snowden.

group, acting as a committee, sent out a call to all the graduates whose addresses were known, for a meeting in Charleston on December 13, 1877.

In the meantime, the committee began the preparation of material for the use of the Association.

Previous to this, application had been made to the War Department at Washington for the return of the Citadel to the State—once in 1869, by Governor Scott, and again in 1876 by Governor Chamberlain. In both cases, the Secretary of War had declined the request on the ground that the Citadel had been "property in an insurgent State" which had belonged to a "hostile organization," "employed in actual hostilities on land," and when occupied by Federal troops on February 18, 1865, became *ipso facto* the property of the United States under a Supreme Court opinion, United States versus Klein.

The committee, consisting of P. F. Stevens, C. S. Gadsden, F. L. Parker, A. H. Mazyck, C. I. Walker, S. B. Pickens, and S. C. Boylston, prepared a complete statement of the question at issue which they submitted to Governor Hampton on August 22, 1877, and was transmitted by him to President Hayes. On October 25, an adverse reply was received from the Secretary of War, closing with the suggestion: "Congress, to which a State can so readily appeal through her Senators, was regarded as the only proper tribunal to entertain such a request."

Such was the status of affairs when the momentous meeting of the graduates was held in the armory of the Washington Light Infantry in Charleston, on December 13, 1877.

At this meeting was inaugurated the movement, seemingly an unpromising adventure, to recover the Citadel from the Federal government, and the still more forlorn hope of getting a bill through the State legislature carrying an appropriation for reopening the school.

As the sequel shows, the first object was accomplished in a surprising manner. The second objective was won only after a remarkable campaign in which victory at the last moment was snatched from the jaws of defeat.

When United States troops entered Charleston, February 18, 1865, it had been evacuated by its defenders, and everyone knew that the war was practically over. No considerable force was needed, therefore, to occupy and hold the city, and as accommodations for troops were not plentiful, the deserted Citadel, with its strategic location offered the most favorable habitation for the troops which were to remain in control.

It is hardly likely that the commanding officer, Colonel Bennett, concerned himself much with the legal aspects of the case when he took possession of the vacated Citadel. He was informed, no doubt, that the corps of cadets was in the field, retreating with Hardee's army into North Caro-

The Citadel during occupation by Federal Troops: 1865-1879

lina, and probably considered himself fortunate to fall heir to quarters so admirably suited for the Federal garrison which would be in control of the city. But later on, when Governor Chamberlain asked for the return of the property of the State, and presented a claim on the Federal government for a rental of $10,000 a year, it became necessary to show some authority for its occupation and use. It would be interesting to know the reasons which influenced Governor Chamberlain to make this demand.

The grounds were formulated by the Post Quartermaster, December 23, 1875, as follows:

"The Citadel, being occupied at the beginning of the rebellion by the corps of State Cadets, the said corps of Cadets marched out of the Citadel as a corps under the command of their professors as officers, joined the forces in rebellion against the United States, and served as an organization for a period of time with the said forces.

"They were succeeded in their occupation of the Citadel by Confederate troops, and the Citadel was held and occupied during the rebellion as a Confederate post, and such occupation terminated only when the Confederate forces were forced to abandon the city of Charleston, immediately after which the city and its Citadel, together with all the forts in the harbor, fell into the possession of the United States, and were occupied by its military forces, and have continued to be so occupied to this date."[6]

As to the claim of the State for rent, the Judge Advocate, in his report to the Secretary of War, Alphonso Taft, March 18, 1878, says:

"So far as this claim is concerned for rent from 1865 to the present time, I consider that it is abundantly satisfied by the care which has been taken of the property and the repairs which have been made upon it by the United States, and I do not think any other payment of any nature whatsoever should be made."

Colonel Hunt's endorsement is to the same effect. He says:

"If the Citadel is hereafter returned to the State, it will be in my opinion an act of pure grace and favor on the part of the government." Furthermore, he thinks the application is inopportune. He continues: "I would add that in my opinion it would be unwise at this period for the government to give up its possession. If at any time in the last ten years it has been important that the government should hold it and keep a garrison on it, this is the time; and the garrison it now contains—a battery of foot artillery serving as infantry and a mounted field battery—is the garrison required."

In view of the unsettled political and economic conditions in the State

---

[6]The contention that the Citadel had ever been the property of the Confederate government, or been occupied by Confederate troops was entirely erroneous.

at this time, it seems remarkable that Governor Chamberlain did not take Colonel Hunt's view of the situation. Possibly he did, and his claim against the United States was a gesture for some local purpose.

When, however, the Hampton administration had become established, and the Federal troops withdrawn from the State, a more determined effort was made by the friends of the Citadel for its return and for a compensation for its use by the Federal troops.

The return of the property without any remuneration for its use and for the accidental destruction of the West wing could at this time have been accomplished without much difficulty. In fact, a bill was introduced in the Senate of the United States on June 4, 1878, by Mr. Randolph, from the Committee on Military Affairs, authorizing and directing the Secretary of War to restore to the constituted authorities of the State the Citadel property with the following provisions:

"Provided, however, that said restoration of said property shall be accepted by the State of South Carolina in full satisfaction and discharge of all claims preferred, or to be preferred, against the United States growing out of the occupation of the same by the United States since its capture."

A petty proceeding, truly, when a great government seeks to shirk responsibility by a provision of this sort.

Fortunately, the Board of Visitors declined to receive the property back on those terms.

There was also a very practical reason for postponement.

"If the building," writes General Hagood, "is returned to us with one Wing destroyed by fire during the U. S. occupation, and no compensation for that and the use of the remainder of the building for 13 years, it will be an elephant on our hands. In the present dilapidated condition of the State finances, the reëstablishment of the School will be out of the question, and we will have for our pains the expense of guarding the building for many years."

The problem, however, was temporarily solved by the War Department. Just as the military authorities had marched in and occupied the Citadel without any right or authorization, so they quietly withdrew without notice or explanation. Early in April, 1879, the Citadel was entirely abandoned by the United States troops. The "white elephant" was left quietly grazing on the Citadel Green waiting for the State keeper to take charge of him.

It became apparent that if the Citadel was to be reopened it would be done without any voluntary aid from the Federal government. On the other hand, General Hagood's opinion, expressed in June, 1878, that a

State appropriation for the purpose was "out of the question," left the alumni with a gloomy prospect of accomplishing their purpose.

But occasionally miracles happen—and one happened now.

At this point it will be convenient to give some account of the reorganization of the Association of Graduates, whose influence must receive the chief credit for the school's revival.

How the Citadel was Reopened

The memorable 13th day of December, 1877, was a busy one for the alumni who had come to Charleston to revive the Association of Graduates and plan for the reopening of the Citadel. Beginning shortly after breakfast, five successive meetings were held before midnight, when the alumni, ex-cadets, and friends of the Citadel retired, weary but satisfied with a good day's work done. At the first meeting, at 10:30 o'clock, held by invitation in the armory of those historic friends of the Citadel, the Washington Light Infantry, only graduates were present. The roll of those present is sufficiently interesting to be preserved:

Class of 1846: John H. Swift,
" 1847: Johnson Hagood, J. P. Southern,
" 1848: J. W. Gregorie,
" 1849: P. F. Stevens, J. B. White, G. B. Lartigue, T. E. Strother,
" 1850: None,
" 1851: J. P. Thomas, N. W. Armstrong, J. J. Lucas, T. H. Cooke,
" 1852: C. S. Gadsden, J. W. Daniels,
" 1853: No class graduated this year,
" 1854: A. H. Mazyck, D. R. Jamison,
" 1855: W. F. Nance, B. B. Smith, D. S. Kirk, F. L. Parker, J. S. Mixson, E. J. White,
" 1856: J. F. Lanneau, H. S. Thompson, R. M. Sims,
" 1857: T. S. Hemingway,
" 1858: No class graduated this year,
" 1859: T. H. Law, T. A. Huguenin,
" 1860: None,
" 1861: C. I. Walker, R. O. Sams, S. B. Pickens, S. C. Boyleston, T. E. Raysor, J. T. Morrison,
" 1862: J. R. Mew,
" 1863: F. M. Farr, A. Doty,
" 1865: J. V. Morrison.

96   THE STORY OF THE CITADEL

Rev. P. F. Stevens was called to the chair, and made an inspiring keynote address, closing with the words:

"The (Citadel) men who had been trained to the bugle, when the clarion sound was made for war, had sprung at once to arms, as the roll of two hundred who served bravely during the War was an evidence. There is scarcely a member who does not bear on his person the scar of a wound received in the heat of battle. Of our two hundred men who served their country, thirty-five have died on the field, and I propose that the names of these martyrs be placed first on our roll, to be called at every meeting, the members present answering, 'Died for his Country.'"

Colonel Thomas had prepared the Roll of Honor, and it was read, the Association standing, and repeating together after the first and last name, "Died for his Country"—a custom which has faithfully been followed in all later years.

Roll of Honor
Died for their Country

Name	District	Class
C. C. Tew	Charleston	1846
J. B. Colding	Barnwell	1851
F. Gendron Palmer	Charleston	1851
D. T. Williams	Beaufort	1852
W. S. Brewster	Charleston	1852
H. B. Houseal[7]		1852
T. W. Fitzgerald	Pickens	1852
R. A. Palmer	York	1852
Micah Jenkins	Colleton	1854
C. T. Haskell	Abbeville	1854
D. G. Fleming	Richland	1854
J. M. Dean	Spartanburg	1855
J. D. Nance	Newberry	1856
J. A. Evans	Georgetown	1856
J. A. Finch	Fairfield	1856
G. Ross	York	1856
C. W. McCreary	Barnwell	1857
W. P. Shooter	Marion	1859
J. L. Litchfield	Horry	1859
O. J. Youmans	Beaufort	1859

---

[7] The name of H. B. Houseal, Class of 1852, Lieutenant Company H, 7th Florida Volunteers, C.S.A., who died in service, was added to the Roll by resolution of the Association, December 15, 1898.

Name	District	Class
G. M. McDowell	Abbeville	1859
T. O. McCaslan	Abbeville	1859
F. H. Harleston	Charleston	1860
A. S. Gaillard	Fairfield	1860
E. A. Erwin	Barnwell	1860
S. S. Kirby	Darlington	1860
F. DeCaradeuc	Barnwell	1860
J. D. Lee	Sumter	1861
J. H. Burns	Kershaw	1861
J. M. Whilden	Charleston	1861
T. M. Wylie	Lancaster	1861
J. C. Palmer	Charleston	1861
N. Wilson	Chester	1861
R. Croft	Greenville	1861
W. F. McKewn	Orangeburg	1862
G. A. McDowell	Charleston	1862
D. P. Campbell	Charleston	1862
G. B. Dyer	Greenville	1862
Wm. Mason Smith	Charleston	1863
J. B. Dotterer	Charleston	1863
J. D. Quattlebaum	Lexington	1864
R. F. Nichols	Sumter	1865

The principal business of this first session was the adoption of a constitution and by-laws for the Association. A committee appointed to prepare these submitted at once the ante-bellum articles, which were being considered when the hour of noon arrived—the time set for the general meeting, at which not only graduates but also ex-cadets, former members of the faculty, and Board of Visitors, officers of the Washington Light Infantry, and others were present.

This meeting was presided over by General James Connor, who had served as chairman of the board for a short while in the closing months of 1865.

The principal business of this meeting was to hear the report of the committee of graduates who, during the preceding months, had been endeavoring to obtain the return of the Citadel by the Federal government. This committee made a comprehensive report, with an exhibit of the correspondence which had been held with the authorities at Washington, and the unfavorable action taken on the request to restore the Citadel to the State and provide remuneration for its use by the United States troops which had occupied it for so many years without legal right.

In order that this matter might be continued and prosecuted by proper authority, a resolution was passed asking Governor Hampton to appoint a Board of Visitors to take charge of Citadel affairs.

Also, a committee of eighteen was appointed to prepare the way by proper publicity for the reopening of the Academy. In connection with this, Colonel Walker announced that plans had been made for the publication of a *Historical Sketch of the South Carolina Military Academy,* which had been prepared by Colonel J. P. Thomas.

At two o'clock the Association reassembled for consideration of the constitution and by-laws, which were adopted after some alterations, and the chair appointed a committee to submit to the Association nominations for officers.

When the Association reassembled at 6:30 P.M., Major Huguenin submitted the following nominations:

For President, General Johnson Hagood; for Vice Presidents, Rev. P. F. Stevens, Major J. B. White, Colonel J. P. Thomas; for corresponding Secretary, Rev. T. H. Law; for Secretary-Treasurer, Colonel C. I. Walker; Directors, G. B. Lartigue, Judge T. H. Cooke, Hugh S. Thompson, Major C. S. Gadsden, A. H. Mazyck.

These were unanimously elected, and General Hagood took the chair which he was to fill until his death in 1898.

Two propositions were discussed and approved by the Association, and were to be pressed with energy: first, the enlistment of all Citadel men in the Association; second, a campaign to "solicit the aid of the press in bringing to the favorable notice of the people generally this call for the restoration of the Citadel Academy to its rightful place in the educational system of the State."

Finally, the exercises of the day were concluded with a social gathering and reunion at 8 P.M.

On April 8, 1878, Governor Hampton appointed the following new Board of Visitors for the Citadel: General Johnson Hagood, Chairman, Rev. S. B. Jones, Colonel Edward Croft, Captain H. A. Gaillard, and Colonel C. I. Walker.

Hereafter, this board was to take responsible charge of the movement to recover the Citadel. But there were to be some years of planning and waiting and doubt.

On February 22, 1879, the Washington Light Infantry dedicated their celebration of Washington's Birthday to the Citadel and the movement for its reopening, and invited Colonel C. I. Walker to be chief marshal of the military parade. There was a meeting of the alumni of the Citadel on that day in the armory of the Washington Light Infantry, but the president of the Association, who was also chairman of the Board of Visitors, could only tell them that the matter of the restoration of the Citadel was still before Congress.

At the annual meeting of the Association on December 11, 1879, the report was still the same. In these waiting years, the alumni over the State were disseminating information about the Citadel, and at the annual meeting on December 9, 1880, held as usual in the armory of the Washington Light Infantry, it was decided to appoint a committee to consult with the Board and with the Citadel men in the legislature with a view of bringing the matter of reopening the school before the legislature then in session. After a full conference, however, it was deemed inexpedient to press the claims of the Academy at this session, as the work of that body had progressed too far.

A memorial from the Association was presented to the legislature December 14, 1880, giving notice that an appropriation for the purpose would be urged at the next session.

The following information was given out by the Board:

"The Board of Visitors have not seen fit heretofore to receive the Citadel building from the United States Government because its surrender has always been coupled with the demand for a waiver of the just claim for rent since August, 1865. The Board felt so long as the State was unable to appropriate funds for the support of the Academy it would not be wise to have the care of the building or to release this claim.

"During the summer of 1880, an arrangement was made with certain agents who agreed to undertake the business of pressing the claim on a contingent fee. The method of working it advised by the agents commends itself to the Board, and they have great hopes of its success.

"By the time the State is ready to meet the question, this claim will have at least assumed such shape that the Board can form a very correct idea of its chances of success—and they believe that at such time the Citadel building can be had without prejudice to the claim; so that when the State makes the proper appropriation, the South Carolina Military Academy can be reopened."

The legislature had been put on notice, and in the summer of 1881 the Association of Graduates started their campaign to win its approval for an appropriation to reopen the Citadel.

The alumni in the various counties were counted on to impress the candidates for the legislature with the merits of the project, and a special committee undertook to enlist the indispensable aid of the press of the State in giving favorable publicity to the movement. Special articles were prepared by influential alumni and furnished to the newspapers with this frank request:

"To the Editor: You will much oblige us by publishing the following article: We are endeavoring to direct public attention to the re-opening of the South Carolina Military Academy as a means of extending to the people the advantages of higher education, and ask your valued assistance." This was signed by C. Irvine Walker, Hugh S. Thompson, C. E. Thomas, and G. B. Lartigue.

These articles developed various phases of the education given at the Citadel. Among the topics assigned for treatment—which did not, however, restrain the writer from wandering into wider fields—were such subjects as these:

"Military Training Useful Principally in the Formation of Character and the Maintenance of Discipline, and Not to Make Professional Soldiers," Colonel J. P. Thomas; "The Educational Results Achieved by the Citadel," Rev. P. F. Stevens; "Some Historical Facts About the Citadel and Its Value to the State," Prof. A. Doty; "Value of the Military School System as a Preparation For Practical Life," Major C. S. Gadsden; "The Movement to Recover the Citadel From the U. S. Government and the Bill Before the Legislature," Colonel C. I. Walker; "The Low Cost of the Education at the Citadel, and the Opportunity For the Poor Boy," S. C. Boyleston; "Results of Military Training on the Bearing, Character, and Spirit of the Cadet," Major J. J. Lucas; "The Influence of the Citadel in the Life of the State," Colonel Coward.

It must not be supposed that all the editors to whom these communications were sent published these articles with approval. A number of them, while expressing every confidence in the gentlemen who, "with pure motives are exerting themselves to resuscitate the grand old institution," were opposed to the movement.

The Anderson *Journal* was disposed at most to receive the information as food for thought. The Hampton *Guardian* was "confirmed in the opinion that it is unfair to the tax-payers to re-open the Academy now." The Abbeville *Medium* even indulges in a tirade to this effect: " 'Has reason fled to brutish beasts that this new imposition must be tamelessly submitted to and no remonstrances made?' "

It can be easily imagined that Colonel Walker and his associates had a strenuous time in the fall of 1881 preparing public opinion to receive favorably the bill to reopen the Academy.

In fact, the situation seemed so doubtful that at a meeting of a group of the alumni in Columbia on November 25, it seemed to most of them that it might be wiser to postpone the Citadel bill to a more propitious time, rather than jeopardize future success by present failure.

The movement, however, had gone too far. Colonel Walker was to have a hearing before the Ways and Means Committee that very day.

He got some reassurance from the favorable report of that committee.

It was not until after the Christmas holidays that the Citadel bill was reached on the calendar of the House.

It came up as a special order on Friday, January 13, 1882, for a second reading. Rather a fateful combination—*Friday, the thirteenth!* Would it be lucky or unlucky?

Here is what happened: Mr. J. C. Wilson moved to strike out the enacting words of the bill. This was tabled by a vote of fifty-one to forty-five. The yeas and nays were requested and recorded. The motion was lost, and the bill was then given its second reading.

Mr. Simpson moved that consideration of the bill be indefinitely postponed, which was lost—yeas, forty-seven, nays, fifty-five. The vote was recorded. So far, so good.

Then came the question: Shall the bill be engrossed and ordered to a third reading?

The roll was called and the vote stood: Yeas, fifty; nays, fifty-five—a disastrous reversal.

So the bill was rejected!

And the House went on with its regular business and special orders until adjournment shortly after two o'clock.

It assuredly looked as if the Citadel bill had been killed. But, luckily, it was not buried. The "parliamentary clincher" had not been put on it, and representative George Johnstone, of Newberry, a valiant and valuable friend of the bill, was recorded as among those who had voted against it. Next day, the 14th, he moved that the House reconsider the vote by which it had refused to order to a third reading the bill to authorize the reopening of the South Carolina Military Academy.

To use a metaphor from pugilism, the bill, "somewhat disfigured," had come back into the ring for the second round.

Mr. Wilson moved to lay Mr. Johnstone's motion on the table. Lost—twenty-nine to seventy-five.

Mr. Simpson moved to strike out the enacting words of the bill. Then some lively scrapping took place. In fact, the round lasted much longer than three minutes.

Finally, however, the roll was called and recorded and showed that the motion was lost by a vote of yeas, forty-four; nays, sixty.

"So the House refused to order the enacting words of the bill to be stricken out," and it was ordered to be engrossed for a third reading.

The Citadel had won its first battle.

A week later, the bill was before the Senate for its second reading, and there was a long and earnest debate on it. Sentiment in favor and against it was so evenly divided, that something very rare in the proceedings of the Senate occurred. On the motion of Mr. Callison to strike out the enacting words of the bill, there was a tie vote— fifteen to fifteen.

In the exercise of his unusual prerogative of casting a vote on legislative matters, the presiding officer of the Senate, Lieutenant Governor John D. Kennedy gave the following reasons for voting "No":

"1st. I favor any scheme which will promote the cause of education in this State.

"2nd. The buildings of this institution having been erected by the State for the purposes of higher education, should be utilized if practicable.

"3rd. I have sufficient faith in the success of the efforts to recover from the General Government the money due the State for the use of the buildings as to believe that the money appropriated by the Legislature will be returned to the State treasury.

"4th. But should this expectation prove groundless, the amount appropriated by the State—one-fourth of which will be paid by the people of Charleston, who most earnestly desire that the institution be re-opened—will amount to so small a sum comparatively that considering the advantages to be derived from it the people will not refuse it. I am unalterably opposed to increasing taxation for any other purpose than education.

"5th. It affords in its beneficiary feature an opportunity for poor, worthy young men to receive a higher education, and they in turn, by teaching two years after graduation, will conduce to elevate the standard of education therein.

"6th. It will be the only really free education institution in the State— the poor boy entering it having his board, clothing, tuition, and all expenses paid while receiving his education.

"7th. The education there received being mathematical, scientific, and practical, the institution in a few years will send forth annually young men who will be trained in those branches which will conduce to develop our material resources."

The Lieutenant Governor had saved the bill.

From this circumstance, the name of General John D. Kennedy has been written in an honored place among the founders of the Citadel.

Many years afterwards—December 14, 1904—Colonel John P. Thomas, presiding at a meeting of the Association of Graduates, referred to the approaching centennial celebration of the South Carolina College in these words:

"In view of the fact that the fathers of the Citadel, Richardson, Jones,

Jamison, and Means were all alumni of the South Carolina College, and in view, too, of the circumstance that it was John D. Kennedy, also an alumnus of the South Carolina College, who gave the casting vote in the Senate of South Carolina that led to the resurrection of the Citadel, after its sleep of seventeen years, I regard it eminently proper for the alumni of the Citadel to send greetings and congratulations to the alumni of the College on the occasion of their coming centennial."

The bill carried an appropriation of $10,000 to put the Citadel buildings in repair, and $5,000 for maintenance for the last quarter of the year.

On March 7, 1882, the Board of Visitors met in Charleston to take possession of the buildings. The transfer from the United States government was made specifically informal, the lone sergeant who had occupied a room in it for some years retiring without orders. But the alumni were not inclined to let the occasion pass without a proper celebration. A considerable number of graduates, ex-cadets, and friends assembled to greet the Board, which was received with an artillery salute fired by the Lafayette Artillery under the command of Sergeant DuBos. The Board was inducted in state into the buildings, and the flag of South Carolina was raised to the top of the flag pole, where it had last flown seventeen years before.

The Board inspected the buildings, arranged for repairs so that the Academy might be opened on October 1 following, and selected the second Tuesday in June for the election of the faculty.

In the evening, in the officers' quarters in the main building, an elegant supper was served, and the following toasts drunk in sparkling champagne:

"The Board of Visitors"; responded to by His Excellency Governor Hagood.

"The Faculty of the Citadel"; General F. W. Capers, former superintendent.

"The Graduates of the Citadel"; Bishop P. F. Stevens.

"The State of South Carolina"; Major H. A. Gaillard, State senator.

"The City of Charleston"; Mayor Courtnay, who concluded his address with these words: "One of Charleston's brightest aspirations will be realized when the cadet reveille is heard once again in the center of our City, and when representatives from every county in South Carolina are welcome guests to the homes of the City by the Sea."

Then followed toasts to "The South Carolina College" and "The College of Charleston," which were responded to by Colonel Chas. H. Simonton and Colonel Chas. Richardson Miles.

A distinguished guest, Dr. A. D. Mayo, of Boston, responded happily to the sentiment: "The Cause of Education," and Captain F. W. Dawson, to "The Press."

Dr. Parker, in offering the tenth toast to "The Militia," paid a tribute to General C. I. Walker as "one who had done more perhaps than any other man for the reëstablishment of our Alma Mater."

In conclusion, Major Huguenin proposed a toast "To the Martyred Dead of the Citadel," which was drunk standing and in silence.

The account of the celebration closes thus: "The hour of midnight was past, and hosts and guests retired, aglow with the thought that the historic building, so long the barracks of the Federal troops, was once more 'The Citadel Academy of South Carolina.'"

THE STORY OF THE CITADEL
PART II

CHAPTER VII
FIRST DECADE: 1882-1892

IF the story of the Citadel during the half century from its reopening in 1882 to this present year, 1932, be divided arbitrarily into five decades, an appropriate title for the first period, 1882-1892, would be "The Struggle for Existence." In this decade the institution was subjected to many and various trials; and that it survived them all may be attributed to the essentially vital idea which lies at the basis of its twofold system of education—sound learning with wholesome discipline.

A catalog of the tribulations which the governing board had to contend with would contain political antagonism of the most serious kind, grave problems of interior discipline, dire poverty, and lesser calamities from the forces of nature, cyclone, earthquake, and fire.

Fortunately, the cadet corps was not oppressed by these problems which caused their elders so much concern, and life at the Citadel, while by no means a bed of roses (nor so intended to be), was an experience that appealed to the sterling qualities of the normal boy.

The act reopening the Academy provided for sixty-eight beneficiary cadets—two from each county in the State at the beginning, but apportioned later according to population. These were selected by competitive examination from poor boys who could not afford to pay for a college education themselves, and were, in general, above the average in intelligence, earnestness of purpose, and willingness to submit to discipline. For many years, it was distinctly these cadets who constituted the backbone of the institution. Statistics for this first period show that sixty-seven per cent of the beneficiaries completed their course, while only twenty-two per cent of the pay cadets remained until graduation.

When the Board of Visitors met on June 13, 1882, to elect three professors for the Citadel, including the superintendent, they could count upon these sixty-eight students at the opening in the fall, but could only speculate as to the number of pay cadets in addition who would apply for admission. About forty-five rooms in barracks were available for cadets after taking out the necessary quarters for the officers, and as these rooms were large, averaging about twenty-two by fifteen feet, four cadets could occupy one without discomfort.

An excellent rooming system was devised, which has been retained with slight modification to the present time. Four folding iron cots of special design could be so compactly stacked against the wall during the day that the space they occupied was negligible. At night, they filled a large part of the floor space. Incidentally, these collapsible cots—folded up each morning and opened out at night— precluded any possibility of infestation by vermin, and though the cadets recited at various times a popular ditty[1] which praised the pertinacity of a certain domestic insect, there is no record of one in barracks during the fifty years of post-bellum Citadel history.

In the daytime the four mattresses were placed horizontally and unfolded, in compartments one above the other, in the lower part of the press —which was the most conspicuous piece of furniture in the room. The upper part of the press had a longitudinal ledge, behind which were two-story compartments in which the underclothes, toilet articles, etc., of the four occupants of the room were arranged according to *Regulations;* and on top of the press, so high that a boy had to tiptoe to reach it with upraised arms, the bedclothes were neatly arranged "with folded edges to the front." As the beds were not allowed down before tattoo, 9:30 P.M., there were many reports against sleepy-headed cadets for "sleeping on ledge of press" or "on top of press" behind the bedclothes.

The uniforms were hung in prescribed order on a line of hooks on the wall, and shoes were arranged on the floor below, heels to the wall. A gun-rack was built on one end of the press with spaces for four rifles, and on the other end was a four-shelf book rack, and also the orderly board, a small wooden plaque containing four slits for the long, narrow, removable cards which bore the names of the cadets living in the room. The one whose name was at the top of the board was orderly of the room—a responsible and unenviable office which had many and onerous duties attached to it, and which lasted for one week. It was assumed in turn by the occupants of the room, and was not looked forward to with any anticipations of delight.

The name of each cadet, in large letters, was also placed over his individual clothes press and clothes hooks, so that responsibility would be fixed for any irregularity in arrangement, and not laid to the account of the orderly.

A metal washstand in the corner of the room was provided with a tin basin, water bucket, and slop bucket. Also, in the early years, before general bathrooms were fitted up, each cadet had a tin bathtub, which hung ordinarily on the nail in the chair rail. Each room was provided with two tables, with drawers at the ends, so that each cadet had a private drawer which could be locked. Four chairs, a trash box, and a blacking stool completed the furniture of the room. Lamps of a cheap kind were furnished, but most of the cadets by permission, provided lamps of their own—the

---

[1]"De June bug got de golden wing,
 De firefly got de flame,
 De bedbug got no wing at all,
 But he gits dar jes' de same."

Rochester Burner and the student lamp, being favorites. One of the daily formations, just before breakfast, was that of the "lamp squad," when the orderly of the room took the lamps to the lamp room—calling for them later in the day when they had been replenished with kerosene.

During the winter season, it was the hard-working orderly who brought up the coal, made the fire, and took down the ashes. It was his duty, also, to empty the slop bucket into the sink on the gallery each morning before inspection and also see that the water bucket was filled at taps as a fire precaution. The fire hazard in those early days was considerable, although the constant presence of the guard prevented a fire getting any headway before detection. Every once in a while, a lamp, turned up too high, would explode, and be carried gingerly to the gallery and thrown like a blazing comet to the quadrangle below. Sometimes, too, the blowers were left too long on the grates, until they became red-hot and ignited the floors near the fireplace. At times, cigaret stumps tossed carelessly into the trash box, started a little excitement, and once the whole lamp room went up into a spectacular blaze, but there was never a serious fire until that of March 14, 1892—and this was not due to any of the causes named above.

On Monday mornings the orderly took the soiled clothes to the laundry room, which was, for many years, merely the room where the colored washerwomen came to get them. It was many years later that a steam laundry with machinery was established. At the end of the week, the orderly carried the fresh clothes up to his room. This hard-working individual was responsible for the general tidiness of his room during his tour of duty, sweeping it out every morning before inspection, and seeing that every article in the room was in its proper place. On Saturday morning, in particular, with the help of his fellows, he had to put the room into spick and span condition for a rigid inspection by the officer-in-charge, and when this was over, he took his name from the top of the orderly board and put it at the bottom with many expressions of relief, devout and otherwise, that his tour was over.

The week each month that a cadet served as orderly was, indeed, a laboratory course in domestic science that had a profound educational effect upon his habits and character, which he could not easily shake off and forget in after life.

With a dormitory capacity of about 175, and an assured enrollment of only the sixty-eight as beneficiary cadets, it may be surmised that the Board's anxiety in the summer of 1882 was that the barracks should be filled rather than any worry about an overflow. As the opening day approached, however, applications for the admission of pay cadets increased in unexpected numbers, and it became apparent that the Citadel would begin its new career with the barracks overcrowded.

It was a motley group of 185 young men who assembled in the quadrangle on October 2, 1882. They ranged in years from sixteen to twenty, and their dress varied from the jeans of the country boy to the Prince Albert coat of the sophisticated young man about town.

At the present time, the recruits who enter the Citadel each fall are provided with uniforms on the opening day, and all distinctions which might arise from difference of dress are obliterated at once, thus promoting the democracy which rests upon merit alone. The new cadet of the fall of 1882 did not receive his first fatigue uniform until November 7. The Springfield rifles were not supplied until the following February 7, so so that the Cadet Corps, when they participated in the exercises incident to the unveiling of the Confederate Monument in Magnolia Cemetery on Thanksgiving Day, November 30, 1882, marched at the head of the procession of military organizations in pea-jackets and without arms.[2]

It was February 22, 1883, that the Battalion made its military debut in full dress under arms. The Washington Light Infantry, by common consent, have long preëmpted this holiday for their annual celebration, and this year, in continuation of their historic friendship for the Citadel, they made the entry of the Cadets into military circles the special feature of their celebration—inviting Colonel Thomas, superintendent of the Academy, to make the address of the day at the Academy of Music.

The reopened Citadel of 1882 was not quite the same as the institution organized in 1842.

In the beginning, the Citadel was a depository for the arms of the State, and its guard was composed of soldiers (boys, it is true), enlisted primarily to protect these munitions and the powder supply kept in the magazines on Charleston Neck.

Governor Richardson's fruitful idea in 1842 was to enlist young men who might profitably spend their time in receiving higher instruction while performing the necessary military duties of soldiers.

In 1882 there were no munitions or magazines to guard, and the students had no necessary duties as soldiers to perform. And yet, the institution was reëstablished with the same strict military system which had characterized it in its earlier years. Colonel Thomas, the superintendent, had been a cadet for four years, in the first decade of the school's history, and had been afterwards both a professor at the Citadel and war-time superintendent of the Arsenal. The discipline of the institution was, therefore, ingrained in his experience, and consequently when he came to reorganize it, this discipline was reincarnated in all its pristine vigor.

---

[2]On every Memorial Day, May 10, since that time, the Cadets have been present at the exercises which are held each year in the shadow of this monument.

The beneficiary feature of the Academy promoted the policy; for, if the State undertook to educate and maintain her indigent sons, she could require that they consider education as a serious matter and submit to the restrictions which were necessary, if they were to make the best use of their time and opportunity. The pay cadets were admitted upon the distinct understanding that they could obtain the same training by subscribing to the same regulations. These were clear in stating that no distinctions other than merit should be made between the two classes of cadets. If the pampered son of wealthy parents found the discipline too irksome—as was often the case—he could get an honorable discharge, but so long as he was a cadet, he had to meet the exacting duties required of all.

It is interesting to speculate what might have been the future course of the Citadel if a man of another type than Colonel Thomas had been called to preside over it in these first formative years. Certainly to Colonel Thomas, reared in the strict school of discipline, "duty was the sublimest word in the English language,"—as he insistently impressed upon the cadets. Idleness, laziness, and waste of time became crimes in the military code. Even a cadet's pleasures must be taken in the time prescribed by schedule. The only evening he might spend away from barracks was Friday, and this was the joyful night that was eagerly looked forward to from the beginning of the week. There was no malingering in the cadet hospital as the week-end approached, and Dr. Parker frankly conceded that Dr. Friday surpassed him in the efficacy of his cures. Those wonderful Friday evenings! If today you should ask any alumnus—even after fifty years—what day of the week still seems to be most promising of pleasure, he will pick out the penultimate.

Dancing was then, as it still is, one of the principal pleasures of the Friday evening leave. There were fewer formal hops in the early days, but in the eighties many groups of cadets and city girls had impromptu Germans on these nights. This was long before the days of the movies, of course, but the theater sometimes offered opportunities for culture which now, alas! in this advanced age are only a pleasing memory. Forty years ago, Sheridan's "Rivals" was presented at the Academy of Music with Joe Jefferson as Bob Acres; Mrs. John Drew as Mrs. Malaprop; Mr. Owen as Sir Anthony, and Louis James as Sir Lucius O'Trigger. Could such a performance be seen today?

The cadets, as a rule, had very limited allowances for pleasures such as these, and when a first-class play came to Charleston, the boys in gray generally stood in line in Market Street, and scrambled up the stairs when the doors were opened to get unreserved seats in the "family circle," or even "buzzard's roost." Today, they cherish imperishable memories of

Booth and Barrett's "Othello"; Sir Henry Irving and Ellen Terry in "The Merchant of Venice"; Jefferson's "Rip Van Winkle," or Emma Abbott's repertoire of light opera.

It was several decades later that leave was granted on Saturday nights also. There were never any recitations on Saturday. After the morning inspection, the cadets had leave in the city until retreat—sundown. Saturday night belonged to the literary societies. The Calliopean Society, formed in 1845, and the Polytechnic, in 1847, were reorganized shortly after the reopening, and performed an admirable service in fostering many a young orator, debater, and essayist. They even introduced a new word into our language—*debatant*.

When, in later years, intercollegiate athletics came to absorb so much of the nervous energy of college students on Saturday afternoons, at which times the contests took place, the corps was too hilarious from victory on the field of sport, or too mentally depressed from defeat to care for the mild and insipid jousts in the arena of letters. In the twentieth century, the major sports of football and baseball preëmpted the week-end afternoons, and basketball and boxing drew great crowds to Alumni Hall on Saturday nights, so that the memory of the literary societies has passed out. Things are done differently now. The Round Table looks after the intercollegiate debates and oratorical contests, *The Bulldog* (college weekly), and *The Sphinx* (college annual), and *The Shako* (college literary quarterly), offer fields for those with literary inclinations. No longer do Calliopeans and Polytechnics scrap on the gallery for the honor of their societies.

There was one feature of college student life which was very popular in the civil institutions of learning, but which was not adapted to a military régime. A number of sub-rosa chapters of Greek-letter fraternities were organized at the Citadel in the eighties, but it was early perceived that their influence was not beneficent. In a small student body, where every individual was well known to every other, and where the relations of the cadets were those of a common family life, any separation into different social groups led inevitably to rivalries and antagonisms, which reacted unfavorably on discipline and the morale of the Corps. The immortal words attributed to Lee, "DUTY IS THE SUBLIMEST WORD IN THE ENGLISH LANGUAGE," were framed and hung on the guard-room wall where every cadet on duty must perforce read them; and undoubtedly it inspired many a cadet officer to perform his military functions with strictness and impartiality. But there were others not so strong, and when an officer-of-the-day was conveniently unobservant of the delinquencies of a fellow fraternity man, while zealous enough to report others, his integrity as an officer was promptly questioned by the corps.

In time, too, there arose a serious division of the cadets into Frats and Non-frats, with intimations of social distinctions which had no place in a democratic body where character, ability, and industry were the only criteria of merit.

The Board of Visitors wisely took measures to abolish and ban the fraternities, an action later approved by the legislature, which enacted a law prohibiting them in all State institutions of learning. This law lately has been repealed, but opinion at the Citadel has not changed.

Some of the features of the Citadel of the early eighties would be considered extraordinary in this present age.

One was the unusual length of the session, which began on October 1 and lasted until the end of July—about six weeks longer than the modern college session.

Another was the inflexibility of the curriculum adopted. Following the custom of ante-bellum days, there were no elective studies for boys of different aptitudes, but all were required to take the same course. In a small institution, such a system seemed necessary for reasons of economy, since elective courses of less than half a dozen students would entail an impossible cost in tuition. It was the general opinion, also, that students of average ability and studious habits could accomplish the required course satisfactorily, and that as a rule boys of sixteen to eighteen years of age who went to college had rather indefinite ideas about their own mental aptitudes, and were not competent to select courses of study for themselves.

The curriculum, therefore, was planned—with one exception—to be broad and cultural, and everyone required to pass up from year to year without conditions. If a student failed in only one subject, that barred him from advancement, and he had to repeat the work of the entire class. It is not surprising, therefore, to read statistics showing a very high percentage of freshman mortality.

The course in mathematics inaugurated by that distinguished teacher, Major William Cain, which was retained for nearly two decades, was remarkable for having a five-hour-a-week schedule for the full four years. It may be of interest to record here the studies, and the time given to this unusual program:

    Robinson's Arithmetic . . . . 10 weeks,
    Wentworth's Algebra . . . . 21 weeks,
    Wentworth's Geometry . . . . 18 weeks,
    Wentworth's Trigonometry . . . . . . 10 weeks,
    Davies' Surveying . . . . 18 weeks,
    Bowser's Analytic Geometry . . . . . . 16 weeks,
    Taylor's Calculus . . . . 24 weeks,

Lanza's Mechanics .... 20 weeks,
Cain's Retaining Walls .... 2 weeks,
Cain's Bridges .... 3 weeks,
Smith's Mechanics .... 13 weeks,
Total number of hours, 775.

This was Cain's Course, and when supplemented by Lyman Hall's allied drawing courses, given two to three hours a week, and including descriptive geometry and various branches of mechanical drawing, it made up a mathematical schedule so unparalleled in the curricula of the colleges of the day, that one educator, when told about it, refused to believe that it was actually given. To his mind, there was "no such animal."

The superintendent, Colonel J. P. Thomas, had been superintendent of the Arsenal Academy during the War, and had subsequently founded the Carolina Military Institute at Charlotte, North Carolina, from which he came to the Citadel in 1882. In addition to his duties as executive officer and disciplinarian, he had also a professor's schedule of teaching. This must have been the most pleasing part of his duties, for he was an ardent lover of letters, and impressed very strongly upon the cadets the delights to be found in his favorite field of learning. In the whole history of the Citadel there is probably no other record of a professor holding classes *before breakfast,* and maintaining, too, the interest of his pupils in the subject —Shakespeare's Plays!

The department of English—or *belles lettres,* as Colonel Thomas loved to call it—included also the courses in history, not separated until many years later. The books in this department included Reed and Kellogg's English Grammar (the great diagram-er), Swinton's Outlines of World's History, both taught by Lieutenant Henry T. Thompson, Anderson's History of England, British Classics, Swinton's Word Analysis, Shakespeare's Plays, Hill's Rhetoric (a delightful textbook), Shaw's History of English Literature, with lectures on American literature by Colonel Thomas, Alden's Intellectual Philosophy, Hickok's Moral Science, lectures on Greek mythology and the philosophy of history, and Story on the Constitution. Lieutenant E. M. Weaver, who had been detailed by the president as Professor of Military Science and Tactics, was called to assist in the literary department, and in 1886 was the instructor of the seniors in Hickok's Moral Science. He left the Citadel in the summer of that year, but in all the vicissitudes of the later years, he never forgot this course which—as he frankly put it—"he took with the Class of 1886." Thirty-two years afterwards, during the World War, when he had become a major general and was Chief of Coast Artillery at Washington, the president of the Citadel called on him on business at his office at the War Department. His first enthusiastic words after the customary exchange of greetings were, "Bond, don't you remember that year at the Citadel when we studied Hickok's Moral Science?"

The list of studies in the department of science covered a wide range, although there was a lack of that laboratory work by the students now considered so essential to the mastery of any scientific subject. The beginnings of effective laboratories were made about the year 1898, from which time on the work in these departments began to assume a greater importance.

The subjects taught in this first decade included Cornell's Physical Geography, Hooker's Physiology, Roscoe's Chemistry, Gage's Physics, Newcomb's Astronomy, Dana's Mineralogy and Lithology, and Leconte's Geology. In the winter of 1885, the Cadets had the rare privilege of attending as guests of the city of Charleston, a series of three lectures by the famous astronomer, Richard A. Proctor, which made a profound impression upon many of the student body. About this time, considerable interest was excited by a church trial of the eminent educator, Dr. James Woodrow, for teaching "evolution," which was considered at that time—and, marvelous to say, considerably later—as heresy. The cadets of 1886, who studied Leconte's textbook on geology, were quite disposed to accept Darwin's theory, but their brilliant young professor, W. G. Brown, also a wise pedagog, taught the facts of geology as distinct from the speculations of philosophy, and avoided the needless use of a word that was anathema to very many Christian people at that time.

In the spring of 1885 a special company of Cadets was organized to compete in the military prize drill at the exposition being held that year in the city of New Orleans. This company, under the inspiring instruction of Captain Lyman Hall and Lieutenant E. M. Weaver, was trained in the evolutions of the company and the manual of arms until the cadets thought there was no possibility of failing to carry off the first prize in the contest among the military college companies.

On the evening of May 8, an exhibition drill was given by the company in the quadrangle of the Citadel, at which the Charleston public added its endorsement of the excellence of their performance, and the Citadel boys, under the command of Captain Lyman Hall and Lieutenant Weaver, and accompanied by their sponsor, Miss Gertrude Gregg (afterwards the wife of Dr. Charles W. Kollock, of Charleston), left for the exposition with their morale at its highest point.

The trip to the exposition was a great educational event to the forty boys who formed the company.

They spent a week in the grounds, rode on probably the first electrically driven railroad cars; had a trip on the Mississippi River on one of the great freight steamers, with smokestacks so huge that one of the cadets thought the *Pocosin* (our Sullivan's Island ferryboat), could be swallowed down one of them; attended a brilliant social event at one of the homes on Carondelet Street, where their own sponsor, Miss Gregg, was one of the belles; and they brought back a beautiful blue silk banner with an embroidered pelican feeding its young, as a trophy of the military drill; but the historian (who was a member of the company) must be veracious and regretfully say that it was not the first prize which we brought home, but the *third*. The Tuscaloosa Cadets (from the University of Alabama), carried off first honors, and deserved them. These were the days of the cadenced manual of arms, and the execution of "fire by file, lying down" by the Tuscaloosa boys was an exhibition of twenty-four human beings operating as a perfect machine.

The demoralization of the Citadel Cadets over their defeat in the company drill vanished, however, when the "Individual Drill for the Best-drilled Cadet in the United States" was won by James Thomas Coleman, of the Citadel. Throughout this gruelling contest, this cadet's arms maneuvered his rifle with the perfection of precision and rhythm, but his body stood like that rock of Gibraltar with which in later life he became associated as an agent of the Prudential Insurance Company. He was the hero we brought back from New Orleans.

The end of the third session of the rejuvenated Citadel which closed on the last day of July, 1885, was marked by an unfortunate crisis in the affairs of the Academy. An issue arose between the superintendent and the cadet officers which reached the stage of having to be reviewed by the Board of Visitors.

The members of the Board were men of affairs, who had fought through a hard campaign to revive the institution, and knew, as the cadets did not, the precarious basis on which the existence of the Academy rested. The uncompromising idealism of youth and the rigidity of military discipline were forces which, if opposed, could wreck the institution. The Board was responsible for saving the situation. At this late date it would be useless and unwise to discuss the issue involved. Colonel Thomas, the superintendent, who resigned, gives in his history of the Academy these extracts from his letter of resignation to the Board:

"While the serious differences of opinion growing out of a grave question of discipline, and some discord, in general between the Board of Visitors and the Superintendent, make my retirement from my office imperative and irrevocable, let me here express the sentiment that moves me of unimpaired devotion to my Alma Mater. Connected for nearly twenty years with the old Academy and the new, and thus bound by strong ties and tender memories, I shall continue to maintain with unswerving loyalty the cause of the Citadel. No longer serving the Academy under the Board, I shall be found in the ranks of the citizenship of the State doing all in my power to magnify South Carolina's School of Arts and Arms, an institution absolutely indispensable to the full development of the mental and moral forces of her youth. No longer the executive of the Board, I shall be its ally and champion in every wise measure of its future administration.

"Allow me to close with the expression, heretofore made three years ago, of my appreciation of the confidence of the Board of Visitors in calling me to the work of reëstablishing this School, of reluming the fires on the academic altar. In my sphere, I have sought, God knows how earnestly, to lay the foundation broad and deep; to make the scholarship of the Academy thorough, accurate, and polished; its soldiership industrious and genuine; its code of ethics high; in fine, to cast it in a mold as lofty as I could fashion. This has been my firm purpose. How far I have effected it, and to what extent I am myself responsible for my unrealized ideal, I shall leave to the public judgment. But this I know and will declare, that I have builded the best I knew how; that I have tried to comprehend my trust and to keep faithful to it with singleness of aim.

"That the Board of Visitors has sought the good of the charge committed to it I have never doubted; and my prayer is that its administration may be blessed with the largest and best results for the Academy and the country."

The resignation of Colonel Thomas as superintendent having been accepted to take effect at once, Lieutenant Weaver was appointed on August 5 to take charge of the buildings during the summer vacation pending the election of a new executive officer.

He was greeted at the beginning of his stewardship with an unpleasant experience with the elements.

Charleston has suffered and survived many notable calamities, both from man and nature. One of a number of disastrous cyclones which have visited the city at widely separated intervals occurred on August 25, 1885.

A reporter for the *News and Courier*, who could evidently look back over the events of many years, described the desolation wreaked by the storm thus: "Even at the end of the bombardment which Charleston endured for two years or more, the streets were not more forlorn or desolate than they were yesterday morning at the height of the cyclone."

The spire of the Citadel Square Baptist Church was toppled over by the storm at seven o'clock in the morning, cutting in twain the three-story residence of Mr. Thomas Dotterer from roof to cellar, exposing the interior of the rooms but, miraculously, killing none of the occupants. The tall spire of the German Lutheran Church on the opposite side of the Square withstood the force of the wind, but the great metal ornament at its summit was grotesquely distorted, and the metal covering of the spire ripped from its surface and flung afar. Great damage was done along the waterfront, and the property loss in the city reached nearly two million dollars. The Citadel withstood the storm's fury with the loss of the tin roof of the chapel and many slates from the roof of the barracks building.

On September 14, 1885, General George D. Johnston, of Alabama, who had served with distinction as a brigadier general in the Confederate Army, was elected superintendent to succeed Colonel Thomas. He came at a critical time in the affairs of the Academy but assumed his duties with a spirit of confidence and energy.

The general economic depression throughout the country following the War of Secession had affected agriculture no less than other industries.

As early as 1871 the farmers began to seek relief through a coöperative organization known as the Grange, which gained wide popularity in the West and South. The Grange idea originated, so it is said,[3] with a clerk in the Federal Bureau of Agriculture, Oliver Hudson Kelley, in 1866. Its purpose was primarily to restore fraternal relations between the North and South, and it was while he was travelling in the South, and was spending some time in Charleston that he evolved the definite plan of the Grange. First organized December 4, 1867, it became immediately popular in the West, but it was not until May 24, 1871, that Ashley Grange, No. 1, was instituted in South Carolina under the leadership of Colonel D. Wyatt Aiken. In 1875, the organization reached its height in the State, with 332 local Granges and a membership of 10,000 farmers. It was in that year that the foundation of an agricultural college was first proposed.

Coöperative buying and other measures adopted by the Grange did not, however, bring the expected relief.

The abolition of slavery in the Southern states had produced not only a disastrous economic condition, but also a social revolution, particularly in rural life, which was to have a profound effect upon their political history.

The ante-bellum seigneurial plantations were no longer possible, and agriculture had to be reorganized upon a totally different basis. The great mass of small farmers were without capital, and led a hand-to-mouth existence under the lien system of obtaining advances.

In the decade from 1882 to 1892, the price of cotton—which was practically the only money crop in South Carolina—never got as high as ten cents a pound. In 1894, it actually reached the low price of five cents.

In consequence, the plight of the farmers became desperate. An immense acreage of agricultural lands was forfeited for nonpayment of taxes, and

---

[3]Article by Easterby in South Carolina Historical Association Magazine.

the rural population was ripe to follow any leader who would promise relief.

Such a leader emerged from among the farmers of Edgefield County.

"Reform" was the catch-word. Not that corruption or malfeasance of office could be charged to the State government since the radicals had been driven out. What was wanted was a reform of the universal poverty; and not knowing just where to launch an attack on this entrenched enemy, the revolution took as its objective the political methods of the old aristocratic oligarchy. This was the movement upon which Ben Tillman rode into local popularity, and which carried him ultimately to the most astonishing heights of political notoriety and fame. In the early eighties, the tirades of the obscure country farmer, who wrote articles to the *News and Courier* advocating a Farmers' College, and ridiculing the Citadel as a "dude factory," were hardly taken seriously, but in a few years this self-called tribune of the people developed the oratory that elevated him above the ranks of the ordinary demagog. The established order of things was getting shaky, and the alumni of the Citadel wisely sought to placate rather than antagonize the looming dictator.

In the spring of 1886, the corps of cadets numbered only 103 members. The senior (First) class was larger than all three of the lower classes combined—the relative numbers being, fifty-five, nine, thirteen, and twenty-six.

The Academy had opened so auspiciously in October, 1882, that there was disappointment and some wonder that it did not continue to grow.

Among the reasons which may be assigned for the diminishing enrollment can be named with confidence the general poverty in the State and the backward condition of the public school system, with a consequent lack of proper preparation for college work on the part of many who applied for admission; but above all it must be ascribed to the reluctance of many of the young men who came to the Citadel to submit to its discipline and the inherent antipathy of adolescent youth to exercising its brain too strenuously. The record of the Class of 1886 is fairly typical of all which have succeeded it. Of the 189 young men who entered the institution in October, 1882, eight per cent were dropped for deficiency in studies, five per cent were dismissed for serious infractions of the regulations, three per cent died, fifty-six per cent abandoned the course voluntarily for various reasons, and twenty-eight per cent remained to the end and received their diplomas. One recalls the parable of the sower, and wonders at the immutable biological laws which exhibit the same inexorable statistics in the twentieth as in the first century.

In March, 1886, the famous evangelists, Moody and Sankey, held a series of public religious meetings in Charleston, and were followed by Sayford and Towner, who held daily revival services in the Citadel Square Baptist Church for two weeks in April.

The Cadets had the opportunity of hearing these evangelists, and many of them were profoundly affected, "ninety out of 108 cadets professing religion."

One immediate result was the organization in the corps of the first Cadet Christian Association, which had its initial meeting of sixty members on April 25, 1886, at which the following officers were elected: president, Cadet O. J. Bond; vice-president, Cadet T. P. Harrison; secretary-treasurer, Cadet J. P. Kinard; organist, Cadet Benj. Munnerlyn. The ladies of Charleston raised a fund of $116 for the purchase of an organ, and the cadet organization thus begun has continued its usefulness in the student body down to the present time.

The influence of these evangelistic services was felt at the Citadel for many years. In the Class of 1888, three of the nine graduates of that year entered at once upon study for the Christian ministry. An amusing incident in this connection occurred a few years later. It was told of a lady in Orangeburg County—it happened that all three of the cadets who became preachers were from that county—who, when asked where she intended to send her son to college, replied, "He wants to be a minister, so I suppose we will send him to the Citadel."

Early in May, 1886, the cadet corps was taken to Savannah to participate in the celebration of the Chatham Artillery Centennial, and was encamped with the other military organizations in Forsyth Park for about a week.

While not eligible to compete with the militia companies in the prize drill, the cadets were invited to give an exhibition drill on the afternoon of May 6, and the battalion of four companies, single rank, received a great ovation, when at the close of their evolutions, the four companies wheeled into line and marched, battalion front, in perfect alignment, down the parade ground, past the grand stand. The majority of these boys were veterans who had drilled together an hour a day for four years, and had attained a perfection of coördinated execution which justly received the enthusiastic approval of the spectators.

A notable incident of the Savannah trip was the special reception accorded the Citadel Cadets by the venerable ex-President of the Confederate States who was a guest at the celebration. This was held on the evening of May 6, at the Exchange. The Citadel sponsor, Miss Virginia Fraser, on the arm of General Johnston, preceded the battalion, advanced to the chair of the President, was introduced to him, and kissed him. Mr. Davis, though aged and infirm, was greatly pleased to see the gray uniforms. When the line of cadets reached him he took the hands of the two at the head of the column—Walker and Kinard—holding them while he made a talk about the heroism of the Citadel men at Fort Sumter.

Commencement Day, July 28, was made a great occasion, not only because it marked the first graduating exercises of the rejuvenated Academy, but also because the alumni, who gathered in Charleston from all parts of the State to rejoice in the celebration, wished also to magnify the event as much as possible in order to forestall the ominous political opposition which was brewing in the State, under the leadership of Tillman and his reform movement.

The exercises lasted three days, and closed with a great banquet at the Charleston Hotel. The Association of Graduates wisely voted to make no pronouncement on the political situation, but to work frankly with both factions for the continued support of the Academy.

Just a few weeks after the closing of the session, when most of the officers and all of the cadets were away on the summer vacation, Charleston was visited by an unusual and startling catastrophe.

The earthquake of August 31, 1886, was the first calamity of the kind to befall a city in the United States. Although not comparable with some similar catastrophes recorded in European history, such as the Lisbon earthquake of 1754, it was an experience terrible by reason of its suddenness and destructive effects, and awful as an exhibition of the mysterious and titanic forces of Nature. It occurred, too, at night—the first shock coming just a few minutes before ten o'clock—and an added terror was created by the fires which immediately broke out among the ruins in the various parts of the city.

The first impulse of the people was naturally to run out of their houses into the open, and to this were due most of the twenty-seven immediate fatalities—the falling chimneys, gable ends, and parapet walls being largely the instruments of death.

Among the large negro population, more excitable and superstitious than the whites, the demoralization was particularly great, and soon all the public squares and open places were resounding with the wailing and praying of delirious groups who believed that the Day of Judgment had arrived. One estimable Charleston lady confessed that she thought that Gabriel's trumpet had sounded. She told her experience afterwards for the amusement of her friends, but at the time it must have been too gruesome for mirth. It seems that some time before the earthquake a gentleman in the city—one of two casual acquaintances of hers—had died, and as sometimes happens, she had got the two confused in her mind, and was under a wrong impression as to which of them had been laid away in Magnolia

Cemetery. On the night of the earthquake, she had gone to bed early, and was sound asleep when the first shock came. Aroused without warning, and being practically thrown out of bed, she ran in her nightdress out into the street, where the turmoil and noise added still further to her mental confusion, and one of the first persons she saw was the gentleman who she had erroneously thought had been consigned to his last resting place in Magnolia. "My natural thought," she said, "was that Judgment Day had come, and that the grave had given up its dead!"

As several shocks of considerable intensity followed during the night, the people were afraid to return to their houses, and for a week or two the Citadel Green presented the appearance of a motley colony of gypsies. Fortunately, the summer weather was favorable to this outdoor life, and as shocks of diminishing intensity continued at intervals for some weeks—and even months—many people thought it safer to live in their yards until their houses could be repaired.

The loss of life finally attributed to the earthquake numbered a little less than a hundred. The damage to property amounted to several millions of dollars, and it is safe to say that there was scarcely a building in the city that did not suffer some injury from the shocks.

The damage to the Citadel buildings was surprisingly small. The solid walls were not perceptibly cracked, nor the large arches around the interior quadrangle. Only the exterior parapets were thrown to the ground, and much of the plaster on the interior walls thrown to the floor. Major Gadsden, the local member of the Board of Visitors, estimated that repairs could be made for $1,500, and obtained from the Carolina Savings Bank and the Loan and Trust Company, a loan for the amount necessary to put the buildings into condition for the opening of the next session. In view of the uncertain, if not precarious, situation as to the legislature's support of the Citadel, the action of the two banks was an act of consideration for the Academy which was recognized and appreciated by the authorities of the institution.

There was one incident of the earthquake which later proved to be of some consequence to the Citadel. The city guardhouse, at the southwest corner of Broad and Meeting Streets, was so badly damaged by the earth tremors that the city was confronted with the necessity of constructing a new building. In this case, the desirability of a change in location was to be taken into consideration, and the Citadel Board was approached by Mayor Courtenay with the proposal that the State cede to the city a site of about ninety by one hundred fifty feet on the King Street end of the Citadel property, on which a new and more centrally situated police station could be erected.

It will be recalled that the West Wing of the Citadel had been burned in 1869 while occupied by United States troops, and that a claim for damages and for rent had been filed against the Government, which was still pending. The fate of this claim was avowedly uncertain, and even if Congress authorized the payment, and the West Wing should be rebuilt, the proposed police station would not occupy any ground to be covered by it. Mayor Courtenay, an undoubted friend of the Citadel, also presented to the Board of Visitors the vision of a possible reversion of the police station to the Citadel some time in the future when the Academy should outgrow its present accomodations. At this time there seemed little likelihood of such an eventuality, but twenty years later this very thing happened, and the State received back the site, paying the city $22,000 for the building on it.

The new police station had hardly been completed and occupied when unexpected success attended the efforts of the South Carolina representatives in Congress to recover rent and damages incurred by the occupation of the Citadel building by United States troops after the War.

On July 3, 1888, the Citadel claim was taken out of the hands of Judge Mackey and Caleb Bouknight, and placed in charge of the South Carolina congressmen, who did such effective work that on November 10 following they could report to the Board that the matter had been adjusted.

Fortunately, the Board was permitted to use about $60,000 of the amount recovered—$77,250. The burnt West Wing was rebuilt at a cost of $22,333, and important repairs in the main building and the East Wing made, besides equipping the academic departments with instruments and apparatus, and laying the foundation for the laboratories of the scientific courses. Two handsome memorial tablets were erected in the rotunda to the graduates and cadets who had given their lives for the Confederacy, a large bronze seal of the State placed over the sally-port, and oil portraits of the four chairmen of the Board of Visitors were painted by Cole and Pelot, of Augusta, Georgia, and hung on the walls of the chapel. When the new session began on October 1, 1889, the Citadel was in excellent repair.

Encampments

As early as 1883, the Board of Visitors had considered the policy of holding a summer encampment of the Corps of Cadets, presumably in the vicinity of Charleston as a rule, and in December of that year made application to the legislature for an appropriation of $1,500 to buy tents and camp equipage, which was not, however, granted. This "military session" was designed to last about a fortnight just after the close of the academic session in July. The time was to be given to the instruction of the cadets in guard duty in camp, field exercises, practice marches, and experience in living in the open.

It was not until the summer of 1889 that the first encampment was held, and there were special reasons for it at that time because it was opportune to make the institution better known in the upcountry to allay political opposition, and to advertise the disciplinary value of the military training given at the Citadel. In this respect it was a repetition of an old and successful experiment made in 1854, when the battalion of cadets was taken on a notable tour through the upper part of the State.

Through the courtesy of Furman University, at Greenville, the campus of that institution was offered for the encampment, and the citizens of the town gave the cadets a cordial welcome.

The results seemed amply to justify the expectations of the Board, and the policy of holding an encampment outside of Charleston was followed for many years, giving the cadets many helpful and interesting experiences.

In the summer of 1890, General Johnston presented to the Board his resignation as superintendent. He had served for five years as the head of the institution, and had administered its affairs with zeal and enthusiasm. The duties of the office were, however, not easy. The superintendent was not only the administrator, but financial officer, head of the faculty, and commandant of cadets. He had to bear the burden of carrying on the institution in the face of political opposition, and with the most meager provision for its needs, as well as teach some of the classes, and attend to all the petty details of awarding demerits for cadet delinquencies. Hence, an excuse may be found for his half-humorous, half-grim remark to one of his faculty when he was leaving, "I would like to come back five years from now and hold a prayer meeting with my successor."

The Board chose, to succeed General Johnston, an experienced educator, an alumnus of the Citadel with a brilliant war record as a soldier under Lee, and a man who possessed in great measure the qualities of a leader and successful administrator. Colonel Asbury Coward entered upon his duties on October 1, 1890, and continued at the head of the Academy for a period of eighteen years.

One of the most important changes made at the beginning of Colonel Coward's administration was the creation of the position of Commandant of Cadets. Up to this time, the Army officer detailed to the Citadel by the War Department was concerned only with the instruction of the cadets in military science and tactics, but it was a logical and important step to place him in charge of the interior discipline of the cadet corps, and make him the Commandant of Cadets, thus relieving the superintendent of one of his most onerous duties. The officer who first filled this position was Lieutenant John A. Towers, Field Artillery, U. S. A., who succeeded Lieutenant Charles A. Cabaniss, Infantry, on October 1, 1890.

THE STORY OF THE CITADEL   125

During these years the Association of Graduates continued active in promoting a sentiment in the State favorable to the military academy, and at its regular annual meeting, held usually in December, at the Citadel, devised many ways of showing the interest of the alumni in all matters that concerned the welfare of their Alma Mater. Illustrations of this may be cited in the exercises on June 1, 1888, when a graduate on behalf of the Association presented very handsome ivory gavels to the two literary societies, and on July 1 of that year, the Association appropriated $134 to buy a set of the ninth edition of the *Encyclopedia Britannica* as a beginning of a library for the cadets.

During the summer of 1890, they were particularly alert during the political upheaval in the State which carried Tillman and the reform faction into power.

For four years Tillman had been lecturing throughout the State as the tribune of the farmers, advocating with great zeal and ability the establishment of a farmers' college. His dream seemed now about to be realized from the Clemson Bequest, and in his first inaugural address he suggested the measures to be adopted for the maintenance of the Agricultural College, which he said could be ready in two years. He recommended to the General Assembly that the South Carolina College, which a few years before had been organized into the State University, revert to its former status as the State college of liberal arts and be properly supported. As for the Citadel—the "dude factory"—he had become reconciled to the spick and span appearance of the young soldiers in their swallow-tail grey coatees and brass buttons, and was gracious enough to say: "At present the Citadel is doing better work in proportion to cost than the University." Perhaps he had begun to feel the responsibility of office, for he added, "There are too few lights in South Carolina for us to put out any of them." However, he was still obsessed with his farmers' college idea, and frankly warned the Citadel supporters that "when Clemson College shall furnish the military training and practical scientific education which can now be obtained only at the Military Academy, that school will have to show cause for its existence as a charity school for military training."

These were ominous words, and when, on the eve of the Ides of March, 1892, the main building of the Citadel, containing all cadet barracks, the administrative offices, and most of the officers' quarters and the classrooms, was swept by fire and rendered uninhabitable, the authorities of the school were gravely concerned for its welfare.

The fire started about ten o'clock in the morning, when all the cadets were at their classes. It originated in the roof—above the third story— supposedly by a spark from a defective chimney which had been injured by the earthquake, and had gained considerable headway before being observed. The city fire department, a most effective organization, had difficulty in fighting the fire on account of its inaccessibility, and practically the whole of the top floor around the Quadrangle, with the roof, was destroyed before any effective means could be taken to subdue the flames. When the fire was finally under control, the two upper floors of the building presented a gruesome picture of charred timbers and blackened walls.

The personal belongings of the cadets were nearly all destroyed—a loss which was generously made up to them later by the kind people of the city. The "Lady Friends of the Cadets" raised $1,100, the city council appropriated $1,500, of which $500 was for the drawing instruments lost by the Cadets.

By great good fortune, it was not found necessary to disband the battalion of cadets. The old Roper Hospital buildings on Queen Street, which had been vacated a few years before (1886), after the earthquake, and could be reconditioned for service, were offered to Colonel Coward for the use of the Academy until the Citadel could be restored. So that by nightfall on this exciting day, the cadets were installed in their new quarters, and—after having heard reveille in their familiar barracks—very unexpectedly listened to taps in strange surroundings.

In three days' time the commissary and classrooms were ready for service at the temporary site, and the exercises of the Academy proceeded without any further interruption until the close of the session on July 9, when, after graduating exercises at the Grand Opera House, the battalion went into camp at Fort Moultrie, Sullivan's Island, for a military session of two weeks.

The fire insurance on the Citadel buildings was carried in the name of the governor—the property belonging to the State, and the Board of Visitors not being an incorporated body.[4] When, therefore, after the fire on March 14, the Board met to take the necessary steps to restore the burned building, the approval of Governor Tillman was required. If he should oppose the rebuilding of the Citadel on the ground that Clemson was in process of construction and was being made ready to take the place of the Citadel in the educational system of the State, the Board would have been in a serious predicament. Fortunately, the governor met the wishes of the Board, and the reconstruction of the building was begun without delay.

When the cadets returned to the city on the ferryboat from Camp Moultrie on Sullivan's Island, on July 22, and passed Marion Square on their way to the depots to depart on their summer vacation, they could

---
[4] The Board was not incorporated until the year 1920.

already see a nearly restored Citadel, and could happily look forward to returning to their own familiar walls when the new session should begin.

And thus, after many trials and vicissitudes, the first decade of the Citadel's history after its restoration came auspiciously to a close.

CHAPTER VIII

SECOND DECADE: 1892-1902

THE second decade, beginning October 1, 1892, opened in newly-built barracks, but with no increase in the size of the cadet corps. The senior (or First) class as it was then called, numbered sixteen men, the junior fifty-four, the sophomore thirty-five, and the number of recruits admitted to the freshman class was thirty-three. The total, 138, was six less than in the preceding year. Of this total, sixty-eight were beneficiary cadets, showing, therefore, practically an even division of the battalion into pay and beneficiary cadets.

This equality, however, did not long exist, and at the end of the session the pay cadets were in a decided minority. The statistics for this ten-year period show that on the opening day in the fall there were present, on the average in all the classes sixty pay cadets and sixty-eight beneficiaries; and at the close of the session in the following June, the corresponding numbers were thirty-nine and sixty-two—showing "casualties" among the pay cadets of thirty-five per cent, and about twelve per cent among the beneficiaries.

This is not difficult to explain. The beneficiaries were selected by competitive examination from among applicants who could not afford to pay for their education, and were usually, in consequence, boys who had a serious purpose to use their opportunities. It is a striking fact that eighty-six per cent of the high-honor men—the first five in each class—during this decade were beneficiaries; and the first-honor men were without exception beneficiary cadets.

The pay cadets were sometimes unruly or backward boys, in whom parents hoped to see some improvement worked by the Citadel training. This could not always be accomplished. It needs no stretching of the imagination to conceive that many of them would get tired of the discipline and quit. Others would leave by special orders for the good of the service. No doubt the beneficiary would sometimes weary of well-doing also—but he had an incentive to endure hardship, and usually he stuck it out. It is a striking fact that of the 204 graduates of this ten-year period, 137, or sixty-seven per cent, were beneficiaries.

In looking, then, for a subtitle to characterize this decade, it seems not inapt to call it "the era of beneficiary supremacy"—since, during this period, this class of cadets took a decided precedence in numbers, scholarship, and discipline. Indeed, but for the beneficiary feature of the school, it is not easy to believe that it could have survived the serious period of

The Citadel in the Nineties

depression which occurred in the last decade of the nineteenth century.[1]

In spite of the subsequent characterization of this period as the Gay Nineties, the country—and the South in particular—was by no means prosperous, or lively in mood. The price of cotton fell to five cents a pound—a figure that did not make the farmer feel hilarious. That the colleges of the State did not grow and flourish was not a matter of wonder; rather, that they continued to exist was a cause for congratulation.

And yet, it must be recorded as a strange fact that during these depressing years, in spite of unfavorable conditions, higher education in the State was to witness the inauguration of two new institutions of learning that were destined to wield a mighty influence in the commonwealth. These were the opening of Winthrop College for Women, at Rock Hill, and of Clemson Agricultural College, in the Piedmont, on the Fort Hill estate, the former home of John C. Calhoun. These institutions were established, one in recognition of the State's obligation to its girls, and the other devoted nominally to the underprivileged sons of the soil. Both had been ardently promoted by Governor Tillman, whose name will always hold deservedly a place of esteem in the annals of these colleges.

The popularity of the new farmers' college—appealing to the largest vocational element of the population, and the offer of education at a price below that which the other colleges could meet, made it difficult for the latter to keep up their enrollment. Low-water mark was reached at the Citadel in 1897, when the enrollment at the opening in October of that year fell to 106—forty-one pay cadets and sixty-five beneficiaries. At the end of the session, the total enrollment was only forty-five—nine pay and thirty-six beneficiaries. But that is another story— the Rebellion of 1898.

The Semicentennial

"The South Carolina Military Academy has completed the fiftieth year of its existence. The half century of its life has embraced the years most eventful in the annals of its mother State. Born when the social organization under which our fathers lived had reached the culmination of its peculiar prosperity, baptized in the blood of revolution, staggering to its feet at the disastrous close of the contest, and resuming its work with inherent vitality and the unconquerable spirit of its blood; again, as of yore, the Academy holds its proud place among the educational institutions of the South.

"Holding fast to all that was good which it imbibed at its mother's breast, broadening its views to meet the requirements of a new departure;

---

[1] A bill to abolish the Citadel, introduced by a representative from Edgefield County, was defeated by a scant majority, forty-nine to forty-six, on January 18, 1896.

availing itself of all the appliances and methods which progress has evolved, it fulfills its purpose. It tenders to the children of its care the priceless gift of a liberal education, tempered to a field of effort, not in the cloister, but in the earnest pursuits of active life, and governed by the high and noble aspirations of the soldier and the gentleman.

"In all the walks of life the sons of the Academy have found their place. As physicians, as lawyers, as engineers, as architects, as merchants, editors, and railway officials, as instructors of youth, as clergymen, as agriculturalists, as officers of the State, and as soldiers they have made their record.

"From the shores of the Pacific, from the distant prairies, from Northern marts, and from all over our fair Southland, they come today, or send loving greetings to their Alma Mater.

> "Alas that some
> 'At their life-blood's noble cost,
> Pay for a battle nobly lost,
> And lie beneath Virginian hills,
> And some by green Atlantic rills,
> Some by the waters of the West,
> Where myriad unknown heroes rest.'

"They, too, in spirit are with us here.

"My friends, the Military Academy welcomes you to her Jubilee in this fair city—a home in which bright eyes and kindly deeds are all she has ever known."

These were the words of General Johnson Hagood, chairman of the Board of Visitors, at the opening of the exercises to celebrate the semicentennial of the Citadel.

December 20, 1892, was the fiftieth anniversary of the passage of the legislative act founding the Citadel, but the date selected for the exercises was February 23, 1893, in order that they might be held in connection with the celebration of Washington's Birthday, and coincided (lacking one day) with the date of the election of the first Citadel faculty, February 24, 1843.

At eleven o'clock in the morning of this day, the Citadel officials, alumni, and guests assembled in the chapel of the Citadel, and a half hour later formed a procession which left Marion Square, with Governor Tillman and General Huguenin in the carriage at the head of the column, and escorted by the Washington Light Infantry and the Corps of Cadets. The line of march was down King Street, through Hasell, and down Meeting Street to the Grand Opera House (which stood on the site now occupied by the Gibbes Art Gallery).

The principal features of the exercises were the semicentennial ode, written for the occasion and read by the author, Major St. James Cummings, professor of English at the Citadel, and the historical address prepared by Colonel Thomas and read by Rev. Dr. Ellison Capers in the absence of the author, who was confined to his bed from an accident.

At the conclusion of the exercises, Colonel Coward, superintendent, presented to the Association of Graduates on behalf of the author, a specially bound and inscribed copy of the *History of the South Carolina Military Academy*, a monumental work which Colonel Thomas had recently completed after many years' labor, and which had just been issued by the press.

The afternoon was given to a reunion of the alumni, and in the evening the celebration closed with a sumptuous dinner at the Charleston Hotel, at the conclusion of which the following toasts were offered with appropriate sentiments. The first toast was:

THE DAY WE CELEBRATE

"Distinguished as the birthday of the most illustrious man known in modern history, it fitly marks in the current year the fiftieth year of the founding of our Alma Mater, whose invaluable influence in developing and conserving the best interests of our State has already rendered her famous."

General Hagood responded as follows:

"The sentiment you have announced, Mr. Chairman, couples the name of Washington with the South Carolina Military Academy. It was an accident that inaugurated the Academy upon a recurrence of the birthday of Washington, but while an accident, it was a happy coincidence. It was he who first brought to the attention of our infant Government the importance of providing for scientific military education, and two days before he died, in a letter to Alexander Hamilton, upon the subject of the organization of West Point, he said, 'It has ever been considered by me an object of primary importance, and I have omitted no proper opportunity of recommending it in my public speeches and otherwise to the attention of the Legislature.'

"Washington knew—and no one in his own observation had better reason to know—that the substratum upon which military success depends is 'the thews and sinews of brave men.' But he also knew that war is an art; it is a science. It requires and uses all the machinery which the most advanced civilization provides, and the teachings of technical schools are essential in any adequate preparation for martial conflict. The lessons of such schools must be learned within their walls. If left to be acquired in the field, all history tells us it is most frequently done amid disaster and defeat. No wise Government, no patriotic statesman, can afford to neglect such schools.

"The poet may dream of a time
  'When the war drum throbs no longer
  And the battle flags are furled,'

but while men are what they are; while injuries can be inflicted or insults offered; while rights are worth maintaining, freedom keeping, or life living, war will be. And in its fiery furnace victor and vanquished, alike purged of the dangerous humors bred in long-continued peace, will renew their manhood. If a people show themselves unable to stand the ordeal, it must yield to the great law of all animate nature—the survival of the fittest. This is not sentiment—it is fact; it is the providence of God. It has been so from the beginning; it will be so to the end.

"While such is the necessity of military training to nations, and such the effect upon them of war—the inevitable, the result of this training and the exercise of its teachings are no less marked upon the individual. Battle is the culmination and paroxysm of war, and skill as well as courage decides the issue; but battle is not all, and in point of time is but a small part of soldier life. Denial of self, devotion to duty, tenacity of purpose, patriotism must be there to sustain the toilsome march, the noisome trench, or the hospital with its long array of cots on which are stretched pallid forms, and the air perhaps tainted with pestilence. It is these high qualities gilded by the valor which shows itself amid 'battle's magnificently stern array,' that make the soldier character the admiration of men and the love of women. It is the seed of these same high virtues, Mr. Chairman, that our Alma Mater sows. Some may fall where there is not much earth; some may fall among thorns; but her history shows that others bring forth fruit, some an hundred fold, some sixty fold, some thirty fold.

"Fifty years of the life of the Military Academy have passed. Few who were present at her birth, or have guided her earlier fortunes, survive. They point to her record, and lovingly commit her destinies to those who come after them—to men in whom devotion to the honor and welfare of South Carolina is an inheritance of the blood."

The second toast was:

THE STATE OF SOUTH CAROLINA

"Wise mother of great institutions! Realizing at an early date in her history that 'all upward impulses come from above,' and that education of the many is always dependent upon the higher education of the few, she has with steadfast wisdom and liberal hand nurtured her higher institutions of learning. May their light ever shine more and more unto the full enlightenment of all her people."

This was responded to by Adjutant General Farley in the absence of Governor Tillman, who had to return to Columbia immediately after the morning exercises.

Captain F. V. Abbott, U. S. Engineers, responded to the third regular toast,

THE UNITED STATES OF AMERICA

"Glorious Union, in whose strength and greatness each constituent State grows great and strong."

The fourth regular toast was read by Colonel Coward, and responded to by Mayor Ficken.

THE CITY OF CHARLESTON

"Intrusted by the State with the home-guardianship of the Military Academy, she has constantly surrounded her ward with the affluence of her refinement, her hospitality, and her tenderest solicitude. Every graduate of the Academy bears with him through life the impress and grateful memories of these ennobling influences."

The fifth toast was announced by the oldest living graduate present, John H. Swift, a member of the first Citadel graduating class in 1846.

THE ASSOCIATION OF GRADUATES

"The representative type of State citizenship. The past guarantees the future."

This was responded to by Rev. Dr. T. H. Law, first honor graduate of the Class of 1859.

The next toast was announced by the president of the Association, General Hagood, and was drunk in silence, the audience standing:

THE DEAD OF THE ROLL OF GRADUATES

"Some have fallen in the paths of peace; many upon the field of battle; but wherever Death claimed them they were found illustrating the teachings of Alma Mater by loyal devotion to State and Duty."

The next toast was proposed by Major J. B. White, war-time superintendent of the Citadel, to

THE NUMEROUS BODY OF EX-CADETS

"Although for various causes they were untimely weaned, yet they imbibed enough of the spirit of Alma Mater to impel them to lives of usefulness, and many to the attainment of distinction."

Colonel Robert Aldrich, an apt example of the class, eloquently responded.

The eighth regular toast was received with great applause.

OUR FRIENDS, THE WASHINGTON LIGHT INFANTRY

"Throughout the century now closing, closely identified with the Battalion of Citadel Cadets. In peace and war, devoted friends, trusted allies. Only gracious memories are recalled for all the years that are past; only joyous hopes spring up for the future which opens up today, in the bonds of a renewed and continuing friendship. *Esto perpetuo.*"

The response was made by ex-Mayor William A. Courtenay, a former captain of the Company:

"Nothing can be more grateful to a speaker than that his subject is acceptable to his hearers. The toast just announced, and so graciously received, conveys this meaning in the highest sense.

"Mr. President, I am here tonight in 'double trust,' as well for the corps in whose ranks my life has been spent as for that kindred corps with whose fortunes the W. L. I. have been linked for all the years since 1843. The one is the same corps in thought, purpose, and achievement as when in the early years of the century, with the defiant cry: 'Remember the Leopard!' it was mustered into the service of South Carolina by the wise Lowndes, the soldierly Cross, the gifted Crafts. The leopard skin on its helmets for eighty-six years proves the identity. It has been a continuous life of usefulness and honor, in peace and war, and when the time for costly sacrifices came, one hundred and fourteen of them laid down their lives for the State. I could say much more—I can not say less for a corps which has given me all its honors, and which will command my best service as long as life lasts.

"The first association of the cadets with the W. L. I. was a funeral escort in which the two commands formed a battalion—at the burial of the Superintendent of the Academy. On every 22d of February afterwards, the cadets escorted the W. L. I. on parade, and the cadet officers were guests of the W. L. I. at dinner. Two important duties were regularly discharged in the early hours of every 23d of February. All clasped hands around the festive board and sang 'Auld Lang Syne,' and then a proper detail was made to accompany the cadet officers home within the grim sally-port at the Citadel.

"Golden days those, when life was young and hope elate! Who shall say that these social happenings were harmful? Some of those who used to sing 'Cram-bam-bu-lee' and 'Here's a Health to Our Friends, God Bless 'Em' have done the State large service, and she knows it. And when, too, the time for sacrifice came who more ready than cadets, in the 'jackets o' gray,' to lay down their lives for South Carolina?

"Fifty years! What a pleasing retrospect in all the manly virtues, in every duty of life! How many youths, without family influence, fortune or favor, have, by the help of a generous State, in this Academy broken through the shadows of obscurity, and by the light of their talents, virtues, and industry have shone upon South Carolina with a cheerful lustre?

"All honor to the noble group who founded this useful State institution!
'They builded better than they knew.'

"All honor to their successors in this great work of the State who in new times and changed conditions, keep alive the educational fires on this home altar!

"May the corps of cadets be ever renewing itself, ever illustrating the virtue and valor of a worthy and ennobling past, keeping step to the old Carolina precepts of honor and duty through all the future years!"

The ninth regular toast was announced in the classic language of Chaucer:

THE LADY FRIENDS OF THE CADETS

"Lo what gentillesse these women have,
If we could know it for our rudenesse!
How busie they be us to keepe and save
Both in hele and also in sicknesse!
And always right sorrie for our distresse
In every manner; thus show thy routhe,
That in them is all goodnesse and trouthe."

This was responded to by Captain John G. Capers, of the Class of 1886, in happy vein.

A telegram was read at this time from Colonel Thomas. A message had been sent to him earlier in the day expressing the sympathy of the Association in his accident, and extending to him "its hearty thanks for his valuable contribution of the address on this semicentennial festival, and the copy of his newly completed History."

Colonel Thomas's reply was greeted with hearty applause: "Extend to the Association of Graduates my appreciation of the thanks and sympathy tendered me. God bless our Alma Mater and all her Sons."

The last regular toast was made in compliment to the poet of the day, Major St. James Cummings, who spoke of the impressiveness of the occasion, and closed with these words: "It is a privilege to come in even at the eleventh hour to share in such a noble undertaking. If the opportunities of the coming years are seized with the same devotion as of old, the future history of the Citadel is certain to be more glorious even than is its past. I feel proud that you have made me your lyric historian; and your praise for my efforts is indeed precious and sweet. You have opened to me the secrets of your council-chamber, and led me to your festal board, and I have forgotten that I was ever a stranger. Henceforth we are brothers in hope. Long live the Citadel!"

※ ※ ※ ※ ※

In May, 1893, Lieutenant John M. Jenkins, 5th U. S. Cavalry, was detailed as the military officer at the Citadel to succeed Lieutenant Towers, who had died on March 23 after a lingering case of tuberculosis.

Lieutenant Jenkins, we think, must have inherited many of the soldierly qualities of his illustrious father, the lamented General Micah Jenkins, who fell at the Wilderness in 1864, for to the cadets he was the *beau sabreur*, the ideal soldier. He went enthusiastically to work with the military department, and with the willing coöperation of the cadets soon had this feature of the Academy in excellent condition. At the end of the session on June 30, Colonel Coward took the corps to Aiken, South Carolina, for a two weeks' encampment, and introduced a new feature into the Citadel Commencement exercises. Before this, the graduation had always been held in Charleston, and only the three lower classes went into camp. But as the Citadel was a State school, Colonel Coward reasoned that it was not inappropriate to have the graduation at the place selected for the encampment, and thus keep the corps intact until the end of the military session.

At Camp Columbus, therefore, held in Aiken, the experienced senior cadet officers were present.

It was the excellent military work of the cadets that suggested to Dr. B. H. Teague, of Aiken, a Confederate veteran who had made an interesting collection of war relics, to present to the Citadel a trophy for the winner of the annual contest for the title of Best Drilled Cadet. Among Dr. Teague's curios was a piece of oak wood that had once been a part of the U. S. S. *Star of the West*. His happy conception was to have a small piece of this wood sawed into the shape of a star and mounted on a gold medal, to be worn for one year by the winner of the competition in the manual of arms.[2]

At Aiken this first drill for the medal was witnessed by the citizens of the town with the greatest interest. It was won by cadet private A. E. Legaré, of the junior class, who, incidentally, thereby won his chevrons as adjutant for the next session. There are now, in this year 1932, forty names inscribed on this much-prized medal.

During the vacation of this year, on August 27 a very severe tropical storm ravaged the coast of South Carolina, creating great damage and loss of life. Fortunately, the damage to the Citadel was not great, and added only $503 to its debt. On October 1, the building was in repair to receive the corps of 144 cadets who reported for duty.

The cadets of the Citadel during the first half-century of its operation were called to their various duties by the sound of drum and fife. It was not until the new century that they were replaced by the bugle for the official military calls. One faithful musician, who died in 1893, will be remembered with appreciation by all the cadets who were at the Citadel before that time. This was Mitchell, the colored fifer, who began his service as a young man in 1847, and was continuously with the cadets until 1861, when he became the body servant of General Maxcy Gregg, and afterwards General Colquitt, whom he served in the field during the period of the War. Mitchell was with the battalion in 1854 when it made its tour of the

---

[2]See list of winners of this medal in Appendix. pp. 230-231.

upcountry, and when the Citadel was reopened in 1882 no employee was more welcomed back than he to resume his former service as musician, a duty which he faithfully performed up to the time of his last illness. As a mark of respect, the cadets attended his funeral services conducted by Rev. D. J. Jenkins, his colored pastor. It was told that the colored military companies in the city wished to give old Mitchell the honors of a military salute at his burial. Not having blank cartridges, the captain of the company was equal to the occasion; he issued ball cartridges to the firing squad, with instructions—in view of the danger of firing into the air—to aim their guns toward the earth into the sides of the grave. And so old Mitchell was buried with military honors—let us hope, unscathed.

It is rare that an inspector, whose business is supposed to be to pick flaws and find fault, gives such unstinted praise as the Citadel and Lieutenant Jenkins received in the report of the Inspector General on June 4, 1894. Under the head of remarks, Lieutenant Colonel Burton reported as follows:

REMARKS

The South Carolina Military Academy is a military school in the best sense of the term, the expenses of which are mainly borne by the State. The military department is modelled after West Point, and closely follows the national institution in all its details, especially with respect to discipline and infantry instruction.

The cadet corps is organized into a battalion of three companies, fully officered. They were exercised in review and inspection, in battalion form, and drilled in the various exercises in company and battalion drill, in close and extended order, the bayonet exercise and signalling with the flag and heliograph.

The discipline, military instruction, bearing and general appearance of cadets, the general care and condition of arms and equipment, and the entire military aspect of the military department of this Academy admits of no comparison to any of the colleges with which I have had experience. It is so superior in all its methods, scope, appointments and its distinctive military features that it must be classed alone, and can only be compared to our National Military Academy. Their limited means does not permit the extended military curriculum that obtains at the West Point Academy, especially respecting advanced theory and practice in ordnance and gunnery and practical instruction in cavalry drill; but in discipline, methods and the practical and theoretical part of an infantry officer's education they follow closely the West Point methods, and are but little inferior in accomplished work. In the set-up, military bearing, cohesion and drill of all kinds in the infantry tactics this battalion equals any organization in the army, and is but little short of that superb excellence generally believed to be possessed by the national cadets.

Though Lieutenant Jenkins has entire control of the discipline and instruction of the corps, and with his facilities should accomplish commendable results, he is, I think, entitled to greater commendation than is due even a good zealous administration of his department. His work has produced results that are far in advance of even the best of the military professors that have fallen within my observation, in this, that he has inculcated fundamental principles rarely touched by civil schools, and he has further insisted upon precise knowledge of first principles and an exact execution of orders from the manual to the extended battalion movements, not observed by me outside of national troops. The College merits the best support the Government can give to the most advanced of the civil institutions where the art of war is taught, for the reason that the nation receives from it results corresponding to its greatest demands.

During this session, Lieutenant Jenkins had introduced a new feature in the practical military work of the cadets—that of a monthly field day in the country. At the end of the year he reported that the plan had been stimulating and beneficial, and that "these exercises form a training that is exceeded by no body of men in the Army."

At the end of the session, it was proposed to stress this feature in connection with the summer encampment by having the cadets exercised in more extended marches, so that they would learn in the best way—according to an old adage—the lessons of security and information and other military subjects from practical experience in the field.

At the close of the academic session, on June 14, the battalion, including the senior class, left Charleston by rail for Rock Hill, from which point they had a two-day hike to Yorkville, a distance of seventeen miles, with a night's stop at Tirzah.

At Yorkville, camp was pitched on the grounds of the former King's Mountain Military School, founded by Micah Jenkins and Asbury Coward in 1855, just after they had graduated from the Citadel.

A break in the routine of Camp Micah Jenkins was a three-day march to the historic battlefield of King's Mountain, where the first vital blow was given to British power during the Revolution on October 7, 1781. On the second day, after the visit to the battlefield, the return march from the mountain was made in a summer thunderstorm that simulated realistically the artillery cannonade of battle (modern battle, of course—there was no artillery at the battle of the Mountain in 1781). The boys splashed onward through the blinding rain and the slippery red York mud, in column of fours, unmindful seemingly of the crashing lightning that smote the mountain.

At Bethany they enjoyed a hot supper, dried out their clothes by the comfortable camp fires, and then called on Colonel Coward to tell them the story of the battle.

It was not always fair weather in camp. On one afternoon a severe windstorm laid the tents flat on the ground. Falling walls in this case fortunately did no damage. The baccalaureate sermon, by Rev. W. R. Atkinson, was delivered on June 24, and on June 29, Commencement exercises were held in the county courthouse, with a splendid address by General John D. Kennedy, who, in his opening remarks, referred to the momentous service he rendered the Citadel in 1882, when, as presiding officer of the State senate, he cast the deciding vote which reopened the institution.

"It was my good fortune once to render a service to the South Carolina Military Academy—a service which I have never regretted, and which time has most amply vindicated. And in the years which have elapsed, as I have watched its progress and success, I have felt for it a strong personal attachment.

"This feeling prompted me to accept the invitation to deliver the Annual Address on this occasion, and I shall be amply repaid if anything which I shall say may conduce to stimulate these young men as they go forth from their Alma Mater to engage in the battle of life to increased devotion to their State and a higher appreciation of their responsibilities as citizens of our common country."

After a forceful discussion of his subject, "The Obligations of Citizenship," he closed with these words to the fifty young men who were about to graduate:

"This great country presents a magnificent field to the enterprising, self-reliant, honest, and sober man. Nor does it matter how crowded the avocation you may select may be—in the language of Daniel Webster, 'there is always room at the top.'

"You are now to step forth into new relations, and assume new responsibilities. Let me urge you to cultivate the virtues of true manhood, love for our common country, and devotion to your State. And in all the years of your lives—of usefulness, honor, and success, I trust—down to a green old age, I know you will never forget your beloved Alma Mater, but, in the words of Scotia's bard,

'The Bridegroom May Forget the Bride   The mother may forget the child
  Was made his wedded wife yestreen,   That smiles sae sweetly on her knee,
The monarch may forget the crown   But I'll remember thee, Glencairn,
  That on his head an hour had been;   And a' that thou hast done for me.' "

Whatever may be the basis for it, the history of South Carolina has been affected by a feeling of *separation,* not to say antagonism, between the geographical sections known as the upcountry and the low country.

This prejudice has sometimes been played upon by politicians, with effects that have been hard to nullify by the best efforts of the broadminded people on both sides of the line. Even in recent times divisions have occurred which make it evident that the era of coöperation has not yet arrived. Two prominent religious denominations in the State have separated into up-country and low-country groups during the twentieth century.

What the geologists call the "fall line" runs across our State in a northeast direction through the counties of Aiken, Lexington, Richland, Kershaw, and Chesterfield. It seems that in some remote time, this line marked the seashore of the Atlantic, and it was the slow washing away of the highlands of the northwest that formed the great alluvial plain of the low country. Navigation at the present time on all the large coastal rivers terminates when this old coastline is reached, where rapids and falls are encountered. It may be, therefore, that the political differences of the people of the State have an origin in geologic causes, absurd as it sounds.

Travel, in the nineties, other than by railroad, was not one of the popular means of recreation. There were no good roads, automobiles, or bridges across the great rivers. And the people were poor. If, therefore, in order to dispel prejudices the people must know each other better, the educational institutions of the State were a potent factor in accomplishing this end.

It is a striking fact that at the end of the nineteenth century the institutions of higher learning in South Carolina were almost wholly confined to the upcountry—Charleston being the only city in the low country with an institution of college grade.

The fact that hundreds of boys and girls from the coast counties were going to colleges in the Piedmont was a promise that gradually a better understanding and sympathy between the sections would result.

Also, when the Citadel was reorganized in 1882, it was well that half of the beneficiaries should come from the other side of the "fall line" to be educated in a city so rich in history and the traditions of culture, refinement, and hospitality as Charleston.

One underlying purpose of the summer encampments of the cadets was to take the Citadel metaphorically to various other parts of the State so that the people might know more about it and see something of its product in its well-trained body of students.

The reverse process of bringing the State—in the representative personality of the General Assembly—to Charleston and the Citadel should be similarly effective. So, in February, 1895, when the legislature was in session, the Citadel and the city jointly sent an invitation to that body to visit Charleston on the holiday incident to the celebration of Washington's Birthday. The invitation was accepted, and a special train brought the solons to the City by the Sea. They were received on Marion Square with military honors, an artillery salute being fired by the cadets, after which there was a review of the battalion and a reception at the Citadel. This was followed by a steamer trip around the harbor, which was a novelty to many of the lawmakers from the interior of the State. After a collation, the guests were present at the always brilliant Washington Birthday military parade, and, weary from a full and busy day, they left the city to return on their special train to the capital.

In the summer of 1895, it was planned to have the battalion go into camp in Columbia for a few days and then take a march of three days to Camden, where the military session would be concluded and the graduating exercises held. While encamped in Columbia on Gadsden Green, near the University, a special review was given in honor of the governor.

The march to Camden was an experience that tried the soldierly endurance of the cadets. In 1895, the road from Columbia to the old town on the Wateree was one of the sandiest in the State. A march of a dozen miles over such a road, with a midsummer sun beaming down upon them, had very little of the element of frolic in it to the cadets. Several streams had to be forded, and dust and water and perspiration combined to make them look very much bedraggled. When, after three days of this trying experience, they dragged their tired feet into Camden, it must be confessed that they looked more like some Confederate soldiers coming from Appomattox than the spick and span battalion of Citadel cadets.

But Camp Kershaw was soon pitched, and the bath tents worked a wonderful transformation, and when the cadets were cleaned up and had on their full-dress uniforms, they looked very attractive to the charming girls who welcomed them to their hospitable town.

At the end of the encampment, June 30, the graduating exercises were held at the courthouse, the Commencement speaker being that accomplished orator, Honorable Leroy F. Youmans, of Columbia. As was expected, the address was resplendent with oratorical effects, rich in classical references, striking in its depth of political knowledge, and also ominous in its analysis of the sinister forces which civilization and liberty always have to fight.

In his closing remarks, he said to the graduates:

"Young gentlemen of the Graduating Class: In ante-bellum days, I well remember a new Northern geography introduced into our schools in which the initial sentence of the chapter devoted to each of the States told for what that State was distinguished. I shall never forget, child as I was, how my pulse quickened and my cheeks flushed with pride as I read that South Carolina was 'distinguished for the opulence and hospitality of her planters, and the number of able and eloquent men she had sent to national councils.'

"That opulence has gone—it may be forever.

"If it can return only *per nefas*—only by 'crooking the pregnant hinges of the knee that thrift may follow fawning'—let it stay gone forever! A true man puts the gifts of fortune to honorable uses, when he has them; but when the fickle goddess shakes her rapid wings, he resigns her gifts, wraps himself in his native worth, and takes honest and undowered poverty for his lot, without lamentation. So said the noble soul of that genial old pagan, Horace. So says every noble soul today—though wealth and office are the only sources of power that are generally acknowledged, and we are strenuously taught, by precept and example, from our cradle up, to clutch at gold and cater for popularity.

"As for hospitality, its seat is in the heart. A man can be as hospitable with his hog and hominy as with Westphalia hams and Tokay wines of comet years from imperial cellars. Francis Marion, the typical knight of South Carolina chivalry, as he appears on the canvas of White, was as hospitable to the British officer with his menu of roasted sweet potatoes as he could have been with soups of bird-nests from the Indian Ocean, or *paté de foie gras* from Delmonico's.

"But if South Carolina is to have a breed of noble bloods, who will make her tower in her pride of place; make an earnest and living reality—a vital principle—of her proud heraldic legend, *animis opibusque parati;* sons who will unite the soaring spirit with sustained strength; sons who will show no degeneracy from the high standard of her great of old, whose character was part of the country's honor, whose ability was a part of the country's strength, who illustrated her history in her heroic age—the days that are past and gone— depends more on her young sons now in the formative period of their character and future than on all other causes combined, which can be contributed to a consummation so devoutly to be wished!"

There is a sentence for young orators to recite, young grammarians to parse, and typographers to punctuate.

During the session of 1895-96, with a view of combining education and relaxation, and of advertising the Citadel, the battalion was taken to the Atlanta Exposition for a stay of several days. Again, a little later, the cadets competed in an interstate competitive drill held in Savannah, winning first prize in the Cadet Class. By invitation of the city of Sumter, the encampment was held in the summer of 1896 in the "Gamecock" town. An educational feature of the stay of the cadets in Camp Kennedy was a trip by rail to the historic battlefield of Eutaw Springs. The Corps was accompanied by General E. W. Moise, of Sumter, who gave the cadets an interesting lecture on the battle, pointing out on the field the position of the opposing forces.

The annual address was made by Senator John L. McLaurin, who had been a cadet under Colonel Thomas, at the Carolina Military Institute at Charlotte, in the seventies. His subject was "The Problems of the Hour," some of which had reference to the Cuban situation, our military unpreparedness, and the value to the country of such institutions as the Citadel. Speaking to the cadets of "the soldier spirit," he amused them greatly by paraphrasing a well-known Biblical parable. He said:

"I have often thought of that prodigal son who left his father's house to see the world (we call it 'seeing the elephant'), and wondered what would have been his fate had he been trained four years by Colonel Thomas or Colonel Coward. (Laughter.)

"He would have seen the elephant—no doubt about that—and possibly 'spent his portion in riotous living.' But when the funds played out, a Citadel boy would never in the world have fed on husks left by the swine. He would have killed the fattest pig in the bunch (laughter), would have eaten spareribs and backbone, and, with the inner man well fortified would —if nothing better turned up—with all the graces born in the City by the Sea have fascinated and married the old man's daughter, and put his father-in-law to tending the hogs. Then with plenty of leisure, he would have entered politics, gone to Congress, paid the mortgage off his father's farm, and provided for the good brother who stayed at home by an appointment to a fourth-class post office. (Laughter.) I know some of the boys who have made almost as wonderful a record as that.

"A Citadel boy would never have gone sneaking home in rags and tatters, whining and shedding crocodile tears. He would have pulled himself together, redeemed his mistakes, and when he went back home, gone with his head up like a man. Or, if he had to become 'as a hired servant,' and fed on husks, he would have gone as far away as possible, changed his name, and not disgraced his family by admitting that he was a miserable failure." (Applause.)

The four-year detail of Lieutenant Jenkins expired in May, 1897, and he was succeeded by Lieutenant John B. McDonald, 10th U. S. Cavalry. The Board of Visitors, at its meeting at the end of the session, paid this tribute to Lieutenant Jenkins:

"Without disparagement to other officers who have held the detail, the

Board expresses its opinion of the unequalled benefit that the detail of Lieut. Jenkins has been to the Academy. To his zeal, enthusiasm, and thorough soldiership is largely due the high standard of instruction and discipline in the Corps, so highly gratifying to its friends, and which has received the high commendation of successive inspecting officers of the General Government."

Major Garlington, who inspected the Citadel on April 9 of this year, and who knew that Lieutenant Jenkins' detail was about to expire, had this to say of him in his official report: "Lieut. Jenkins has done most excellent work here, and has demonstrated that he possesses a remarkable quality for training young men. His services would be of the highest value in the tactical department of the United States Military Academy."

The War Department at this time was observing with attention the character of the military work being done in the various colleges of the country where such instruction was being given. Perhaps premonitions of the War with Spain were arousing serious consideration of military preparedness. Under date of July 3, 1897, Inspector General Breckenridge addressed a communication to the presidents of the military colleges expressing the interest which the Secretary of War was taking in their military work, and suggesting that their governing boards give their influence through their senators and representatives in Congress to secure legislation which would "more adequately recognize and promote the excellent and faithful work already shown. . . . Useful in times of peace, such resources would be invaluable in providing our armies with an adequate number of properly educated and intelligent officers in time of war." The communication concluded with the following words, of particular interest to many Citadel men who would have liked an Army career: "Perhaps an additional incentive to stimulate the Cadets under instruction would be secured, with little or no additional expense, by the passage of an Act by some future Congress insuring a commission, under proper competition, to a definite number of Cadets, which may be increased when the exigencies of the service will permit."

Up to the time of the Spanish-American War, any Citadel graduate ambitious for a career in the Army had open to him only the uninviting road of enlistment as a private with the hope that an opportunity would occur later for him to get a commission "from the ranks." A number of Citadel men during the nineties did this, and are now, in this year 1932, high-ranking colonels in the Regular Army. In 1903, a new way was opened to them by the War Department as suggested by the Secretary of War, and a considerable number have been added to the list of Citadel graduates in the Army—as well as the Marine Corps and the paymaster service of the

Navy—by virtue of the Citadel being classed as a distinguished college.

The annual encampment in the summer of 1897 was held at Anderson, South Carolina, June 16-July 1. By courtesy of Colonel John B. Patrick, a Citadel alumnus of the Class of 1855, and his son, Captain Patrick, Camp Calhoun was pitched on the campus of the Patrick Military Institute—another "daughter of the Citadel." Besides the usual exercises and duties incident to the regular routine, a march of forty miles, covering three days, was made to Clemson College and return, stopping at Pendleton both going and returning. At Clemson, an incident of interest was a special address to the two corps of cadets made by Colonel J. P. Thomas, who selected for his topic the appropriate theme, "Shakespeare in Arms."

In December, 1897, the city council adopted an ordinance to support five beneficiary cadets at the Citadel, at a cost of $1500 annually. The number was later increased to six, the allowance for each being reduced to $250 a session.

The year 1898 was eventful in many ways at the Citadel.

At the beginning of the year—January 4th—occurred the death of General Johnson Hagood, chairman of the Board of Visitors since 1877, first honor graduate of the Class of 1847, brigadier general in the Confederate army with a distinguished war record, and governor of the State.

As a mark of respect, the entire corps of Cadets was taken to his funeral at Barnwell, South Carolina. General Hagood was succeeded as chairman by Major C. S. Gadsden, of Charleston.

On February 15 the blowing up of the U. S. S. *Maine* in Havana harbor brought to a climax the growing friction between the United States and Spain, caused by conditions in Cuba. The patience of the American people with the intolerable situation in that island as a result of Spanish misrule was exhausted, and when the Board appointed to investigate the cause of the sinking of the *Maine,* found that it was from an outside explosion, fixing by implication in the popular mind the responsibility upon the Spanish authorities, it became evident that in spite of President McKinley's efforts to avert war the country was contrary minded. An old Texan slogan was revived with a change of name, and "Remember the Maine" became catchwords to fire the nation. The American people were under the impression that Spain was an effete nation, still living in imagination in the glories of the fifteenth century; and the Spanish press incited its public to believe that the Americans were a race of hogs, without courage, and bent only on the pursuit of the Almighty Dollar.[8]

---

[8] One day, during the summer of 1898, the writer was taking a woodland walk with a cultivated old gentleman who was widely traveled and had been many years before an American official at Manila. A pig came out of the woods some distance ahead of us and stopped in surprise to survey us. We had been talking about the war, and the

The War Department was concerned with the practical question whether the Spanish navy and army were not better equipped and more formidable than our own. The Spanish battle ship *Vizcaya* had paid a friendly visit to New York harbor some months before, and the public was in doubt whether any vessel in the United States Navy was her equal in fighting ability. The size of the Spanish army greatly exceeded ours and its equipment was said to be more modern, their Mauser rifle, in particular, being superior to our Springfield. The morale of the Spanish soldier was also thought to be high. In a conflict between numerically equal forces there was speculation as to what the result would be.

Another matter that caused concern to our War Department was the defense of the American seacoast cities if an attack were made by a formidable Spanish fleet.

At Charleston there were some modern emplacements armed with 10-inch and 6-inch disappearing rifles, but immediate plans were made for the construction on Fort Sumter of an emplacement for two 12-inch guns, and for a 10-inch battery on St. Helena Island for the protection of Port Royal.[4]

It is interesting to note (a fact learned after the war), that the orders received by Admiral Cervera when he sailed from Spain with the flower of the Spanish navy, gave him discretionary instructions for an attack on the American coast, Charleston being mentioned specifically as an objective. Thus, one might-have-been chapter in the annals of this old historic city was not written, and can only be speculated upon.

In the midst of all the war talk, a wretched affair occurred at the Citadel which seriously threatened the life of the institution.

No young man who has been one year a cadet can be ignorant of the essential features of a military school. At the civilian colleges, there may be forms of student government to take care of the discipline; but in a military college, every one, from the president down, is subject to clear-cut regulations.

"Par. 57.—Obedience to orders lies at the basis of all military discipline, and cadets will obey without argument or comment all orders received by them from proper authority. If, in their judgment, such orders have been given them improperly, they may afterwards make an official statement of the circumstances in writing to the commandant of cadets, who will investigate and take proper action."

---

old gentleman, on seeing the animal, halted instantly, raised his hat courteously, and said: "Good morning, fellow countryman!" Such consideration being beyond the porcine comprehension, the pig gave a grunt and scampered into the bushes.

[4] The writer recalls with interest that he was employed for five hours a day during the three months of April, May, and June in the office of the United States Engineers at Charleston as a draughtsman on the plans for these works.

Such is the regulation in the Blue Book, a copy of which hangs in every cadet room.

Another regulation which safeguards the personal rights of every cadet reads as follows:

"Par. 58.—If any cadet shall consider himself wronged by another cadet or by an officer of the college, he may submit a complaint to the president, who is hereby required to examine into the case and correct the wrong complained of if an injustice has been done. Should the complaining party be refused redress, he may appeal through the proper channels to the chairman of the Board of Visitors."

The troubles in discipline that have arisen in military institutions have usually been due to the fact that the students who have felt that they had a grievance have been unwilling to submit their causes to the judgment of those in authority over them. They prefer agitation among their fellows, and endeavor to get a combination of the student body sufficiently strong to intimidate the authorities. A strong malcontent can always attract a following of other malcontents, and if they have a plausible cause it is often possible to get those who have no feeling of a grievance to lend their support to it.

"Par. 60 (m)—Combining under any pretext whatever with other cadets or entering into any agreement with other cadets, either in writing or verbally, with a view of procuring a redress of grievances, or violating or evading any regulation of the college, or joining with others in expressing disapprobation or censure of an officer or cadet, or in doing any act contrary to the rules of good order and subordination is an offense of a serious nature which may be punished by dismissal, or expulsion, or otherwise."

The inherent right that men have to associate themselves together for their mutual welfare does not excuse a soldier for insubordination. A cadet is bound by his matriculation promise to play the game according to the rules—and he always has the privilege of resigning. Even if the subject matter of a petition be entirely commendable, it should not come to the commanding officer as the demand of the soldiers. Just after the battle of Waterloo, Lady Shelley asked Wellington why he did not seem pleased at the cheering of his soldiers. He replied: "If you allow soldiers to express an opinion, they may on some other occasion hiss instead of cheer."

Confessedly, Wellington was not a democrat. But the military principle he uttered was sound. No one is worthy to command whose ear is turned to catch the applause or censure of those under his authority.

Of the sixty-four cadets who were dismissed by the Board of Visitors on April 8, for refusing to obey the orders of the superintendent, and for having entered into a combination in violation of the regulations, there were some no doubt who were guilty only on the second count, and suffered with those others with whose fortunes they had unadvisedly tied their own. The Board considerately decided that those members of the two lower classes who were dismissed might be readmitted at the beginning of the next session upon certain conditions, but that in the case of the members of the two higher classes, for whom no excuse of immaturity or inexperience could be offered, their action was final. Twenty-four of the twenty-nine members of the senior class, therefore, lost their diplomas within three months of graduation.

On April 22, Congress declared that a state of war existed between the United States and the kingdom of Spain, and immediately the President called upon the states to furnish their quotas according to population for a volunteer army of 125,000 men.

South Carolina's contribution to the military forces consisted of two regiments of infantry, one company of heavy artillery, one company of engineers, and a contingent of naval militia.

A large number of Citadel alumni immediately sent in their names to the governor, offering their services as officers. It was soon apparent, however, that only a limited number would be needed.

The first military unit in the State to be mustered in was the Heavy Battery of artillery under the command of Captain Edward Anderson, Class of 1886. In less than ten days after the President's call, Captain Anderson had raised the 166 men required for his battery, and reported at the State rendezvous for enlistment. Here, however, more than half of the applicants were rejected on the physical examination, and more than three hundred were examined before the needed complement was finally secured, and the battery mustered in on May 21.

The 1st Regiment of Infantry was mustered in on June 2. In command of that regiment was that admirable officer, Colonel Joseph K. Alston, who had received his military training under Colonel Coward at York, and Colonel Thomas at Charlotte, and then at the V. M. I., where he graduated as a senior cadet captain in the corps of cadets. The Citadel graduates in the 1st Regiment were Majors J. H. Earle and J. D. Frost, Captains W. S. Langford, and L. S. Carson, and Lieutenants C. B. Smith and B. D. Wilson. There were a number of ex-cadets in the regiment also, three of the sergeants being boys who had been in the recent trouble at the Citadel.

As soon as war was declared, events succeeded with astonishing rapidity.

On May 1, Dewey won his incredible victory at Manila, annihilating the inferior Spanish fleet under Montojo without a single casualty on the American ships. Two months later, the pride of the Spanish navy was utterly destroyed in an equally astounding one-sided battle at Santiago.

THE STORY OF THE CITADEL   149

The defeat of the Spanish army, however, was achieved less quickly, and not without sacrifice. The War Department proceeded to dispatch to Cuba and Porto Rico the necessary forces for their conquest.

In Charleston, early in July, five regiments were assembled to take ship for the war zone. On July 7, the 3d Wisconsin Regiment gave a dress parade on Marion Square previous to their departure.

The battles around Santiago took place early in July, and the protocol which stopped hostilities was signed on August 12.

The war was over so quickly that none of the South Carolina troops ever got into action. The 1st Regiment, mustered in June 2, got to Jacksonville, Florida on July 28, and was there when the protocol was signed. The Heavy Battery was stationed at Fort Moultrie until the close of the war.

The 2d Regiment, under Colonel Wilie Jones, was not mustered in until August 23—eleven days after the protocol—but eventually got to Cuba as a part of the army of occupation. The Citadel graduates in this regiment were Major Havelock Eaves, Captain W. C. Davis, and Lieutenants T. C. Stone and E. R. Tompkins.

Three Citadel graduates, civil engineers F. C. Black, S. C. Gwynn, and R. E. Boggs, were lieutenants in the 3d U. S. Volunteer Engineers, which performed useful service in the vicinity of Cienfuegos, Cuba, after hostilities were over.

This regiment, under the command of Colonel David D. Gaillard, of the Regular Army, performed a unique military service before going to Cuba. At Macon, Georgia, it was called out one night about midnight by orders from General Wilson to disarm a refractory regiment of colored troops, the Sixth Virginia Infantry. In twenty minutes, the regiment was ready and performed its assigned duty, furnishing "possibly the first instance in the history of the government of the disarming of United States troops by United States troops."[5] Lieutenant Black was in command of Company H, and took charge of the prisoners.

Besides the seventeen Citadel graduates in the volunteer forces, there were five others in the Regular Army, one of whom, Captain W. H. Simons, was in the fighting to capture Santiago, and was wounded in the arm.

When the session of the Citadel closed on June 28, there were forty cadets left on the roster. In order to help rebuild the enrollment, the Association of Graduates contributed $220 to send out a canvasser for recruits, and the new session opened in the fall with fifty-nine new cadets and a total enrollment of 111.

All army officers on detail to educational institutions having been recalled

---

[5] *South Carolina in the Spanish-American War*, by Adjutant General J. W. Floyd.

during the war, the Citadel had to send for one of her own graduates to act as commandant, and found in Lieutenant J. Willis Cantey, who had graduated as senior cadet captain of the corps in 1893, an officer in every way qualified for the duty. The new session, in spite of diminished numbers and reduced finances, began hopefully. The vindication of the school's discipline had restored the confidence of the authorities and the tone of the corps. It now had the spirit to go forward.

At the end of the session, it was decided to hold the military encampment as usual, and Orangeburg was selected as the place. The Corps arrived at Camp Jamison on Saturday, June 16, in an unseasonable cold rain, and accepted the hospitality of Colonel Clarence J. Owens to make use of the weather-proof barracks of his Collegiate Institute until the following Monday. The military exercises included two practice marches of six or seven miles, and the graduating exercises were held on the 30th, with Senator Tillman as the speaker. He spoke, as usual, without notes, and his address was characteristic—varying from the serious to the facetious and humorous. He recalled that he had once called the Citadel a "dude factory," and turning to the ladies of the audience, he feigned an interest in solving the problem of the attraction brass buttons have for the female sex. From this he proceeded to expound a wholesome lesson on conceit, the sin against common sense, in which he told this story:

"A certain first-honor young lady graduate thought she knew it all. She was shocked that her minister pronounced the word 'wound'— meaning an injury in battle—as if it were the past tense of the verb, to wind. 'Why do you say wownd instead of woond?' she asked. On the spur of the moment, the minister wittily replied that he 'had not *foond* any *groond* for any other *soond!*' "

The session of 1899-1900 opened with 123 cadets—this being an increase of twelve over the enrollment of the previous year. Early in the session, a special committee of the Board of Visitors was appointed to look into the finances of the Academy. The committee made the following report:

"We regret to have to report that the expenditures of the Institution are in excess of its income, and have been for years past. Commencing in 1890 with a deficit of about $1,500, it has gradually increased with varying sums until it now has reached $6,323.64.

"This condition may be ascribed in part to the disastrous fire which occurred in 1892, two storms which worked injuries to the building, the higher prices for all things bought, and the rebellion of 1898, whereby one-fourth of the income of the Academy was cut off without any corresponding reduction in expenses.

"But these events do not account entirely for this indebtedness, and your Committee are forced to the conclusion, reluctantly, but nevertheless certainly, that the support allowed by the State is not adequate to the needs of the institution.

"We have sought but have not found any department of the school which can have a reduction of expenses without impairing the usefulness of the Academy.

"Everything everywhere is conducted with a meagerness that is pathetic, and the fact forces itself upon our minds that if the Citadel is to live, application must be made to the Legislature to come to its rescue, and provide for the extinguishment of this debt, and hereafter to make an annual appropriation more in proportion to the necessary wants of the Institution; and we recommend this course.

"We are satisfied, and feel it due to the Quartermaster, Mr. White, to say, that he is a faithful and efficient officer, and the wonder is that with the burdens he has to carry and the limited means at hand for discharging them the conditions are not worse.

(Signed by) ROBT. ALDRICH, Chairman,
J. J. LUCAS,
COLE L. BLEASE."

It is gratifying to note that the legislature appropriated the funds necessary to liquidate the indebtedness.

A communication from the faculty of far-reaching consequence was submitted to the Board recommending that the degree of bachelor of science be awarded to graduates of the Citadel. Statistics taken from a paper prepared by the president of the University of Tennessee were cited to show how favorably our institution compared with others in its curriculum.

Table showing number of hours in four-year scientific course:
Vanderbilt, 2,280.
Tennessee, 2,800.
Alabama, 2,052.
The Citadel, 3,220 (exclusive of military subjects).

In February, 1900, an act was passed granting authority to the Board to grant the bachelor of science degree to the graduates of the Citadel. The wording of the act conferred a limited power; but at the time it was all that the faculty or the Board desired. Later, it was found desirable to confer other degrees; in fact, half a dozen bites were taken at the cherry in the succeeding years, until at length in 1932 the Board was given "authority to grant degrees"—no limitations being stated.

As showing the ultra conservatism of the faculty, the bachelor of science degree was conferred at first only upon those members of the graduating class who attained an average of eighty-five per cent or better in their four-year course. This was a reward for only those who won very high honors. Only five out of the nineteen members of the Class of 1900 got it. Next year, the number was less—three out of fifteen; and in 1902, the percentage still dropped—five out of thirty-six receiving crowns (or halos). It is not to be wondered at that eighty per cent of the graduates of those years who received their hard-earned diplomas without any degree thought themselves badly treated. Very wisely, the faculty perceived its error, and in 1905, by resolution of the Board, the degree of bachelor of science was to be conferred on all graduates who received the diploma—this action being made retroactive back to include the Class of 1886.

A change of considerable importance was made by the legislature of 1900 in the method of appointment of members of the boards of trustees of the State colleges, The University, the Citadel, and Winthrop. Up to this time, the vacancies in the membership of these boards had been filled by appointments made by the governor. The act of March 2, 1900, made the members elective by the legislature. In the case of the Citadel, the act specified that the "regular members," as distinguished from the ex-officio (State officer) members, should be "graduates of the Citadel." The term of office was fixed at six years.

In May, 1900, the path of a total solar eclipse passed through the Southern states.

The subject of astronomy had been in the curriculum of the Citadel from the beginning.

The writer has before him a copy of the examination questions in astronomy given to the graduating class of the Citadel in April, 1859. It was published in the *Courier* the next day (April 7).

"Professor J. B. White conducted the examination in astronomy.
1. Refraction and its effects. Cadet Law.
2. Parallax. Moon's parallax and distance. Cadet Smith.
3. The sun's apparent path and motion. Cadet W. P. Shooter.
4. Precession of the equinoxes; the cause and effect. Cadet J. E. Spears.
5. Solar orbit an ellipse with the earth at one focus. Time. Cadet W. Adams.
6. Solar spots. Sun's parallax from transit of Venus. Cadet McDowell.
7. Parallelism of earth's axis. The seasons. Cadet T. A. Huguenin.
8. Comets. Cadet O. J. Youmans.
9. The moon, its phases, etc. Cadet McCaslan.
10. Eclipses of the moon. Cadet T. J. Weatherly.
11. Eclipses of the sun. Breadth of shadow at the earth. Cadet P. S. Layton.

12. Different kinds of eclipses, and number in a year. Cadet W. R. Marshall.
13. Fixed stars. Cadet W. Cothran.
14. Terrestrial longitude. Cadet F. L. Garvin.
15. Tides. Cadet Litchfield.

The examinations will be continued today, and at 5 P.M. there will be an artillery drill."

There is nothing remarkable about the questions. They are such as would be given in an examination on mathematical astronomy today. The circumstance that they were published in the daily press, however, leads one to wonder if there could have been a popular interest in so technical a subject. Most probably not. On the day before, the *Courier* had covered the examination on Calhoun's "Disquisition on Government" with the same detail. But in this case the reading public was more intelligently interested—having been educated in its discussions for a quarter of a century.

The historian, in looking over this old examination paper—and who sees the events of the subsequent years—is impressed chiefly, not with the questions, but with the personality of the boys who were standing the test. He sees that in half a dozen years they were all to be tried in the furnace of a great war. Cadet Shooter was to become Colonel Shooter, and fall in battle in Virginia. So with Youmans, Litchfield, McDowell, and Mc-Caslan—on other fields. Tom Huguenin was to become one of the heroes of Fort Sumter; and all the others were to give their service and their blood for the State.

Cadet Layton, in his discussion of eclipses of the sun probably explained that the moon, in circling the earth, comes between us and the sun once a month, and sometimes its shadow cone happens to be long enough and in the proper direction to reach the earth's surface; in which case the fortunate person who chances to be just at the right spot may see the wonderful spectacle of a total solar eclipse. Due to the moon's motion, this shadow spot—a circle which varies in size from nothing to about 200 miles in diameter—moves very rapidly (about a thousand miles per hour) over the surface of the earth, and its path is very accurately calculated in advance and shown on a map so that persons interested may travel to some point where they can see this rare and awe-inspiring sight.

On account of its educational value, the First Class of 1900 made an expedition to North Carolina to view the eclipse of May 28 of that year.

Wadesboro was selected as the point, not only on account of its accessibility, but also because scientific expeditions from the Smithsonian Institution, the Yerkes Observatory, Princeton, and other universities, had selected it for their observations.

On the day before the eclipse, the cadets made the rounds of the various field observatories, looked at the sun's prominences through the Princeton telescope, were given the rare privilege of a talk by Professor Langley, of the Smithsonian party, and in the evening heard the public lectures given by Professor Young and other noted astronomers.

The totality at Wadesboro occurred shortly after nine o'clock in the morning, and lasted about a minute and a half. The Citadel party was stationed near the silk mill, and the members assigned to different observations—some to make drawings of the corona in the four quadrants, others to observe the shadow bands, and a professional photographer to make photographs. These, by the way, came very near being lost when, after the eclipse was over, the photographer mounted with his tripod into the buggy which was to take him back to town. Two sunrises in one morning were probably too much for the comprehension of the country horse. Anyway, the frantic animal made a perilous and exciting runaway. The photographer could do nothing but hold desperately to the buggy and his camera, although every moment it looked as if he would be catapulted into the adjacent fields. Fortunately, the horse kept to the road—a long, yellow road with a heavy upgrade. It was this that averted a catastrophe. Gravitation, another of the forces of Nature beyond the equine understanding, finally conquered the force and endurance of the animal.

It may be of interest to future cadets to know that while any particular city might have to wait thousands of years to witness the phenomenon of a total eclipse of the sun within its corporate limits, there is a calculated prediction that Charleston may have that distinction on March 7, 1970, when, let us hope, the Citadel cadets will make due plans to observe it.

No encampment was held this summer, the Board deciding to send the Corps to the State Fair in the fall instead. The Commencement exercises were held therefore in Charleston, Rev. George H. Cornelson, of the Class of 1888, preaching the baccalaureate sermon on June 24, and the Honorable Stanyarne Wilson delivering the address on June 27. Next year, however, the custom was resumed of holding the graduating exercises away from home, the Corps being taken first to Florence, where they spent the first night, June 14, in a torrential downpour of rain, some of the tenters being obliged to take shelter in the schoolhouse.

The camp at Florence was named in honor of General W. W. Harllee, a prominent citizen of the Pee Dee section, for whose daughter the town had been named.

The cadets broke camp at Florence on June 18, and had a hike of fourteen miles to Darlington, where they pitched their tents on the grounds of St. John's Academy. Besides the usual military duties and exercises incident to the camp, the surveying class ran a hypothetical trolley line from the town to Mineral Spring, three miles distant. Also, the class in astronomy had previously calculated and constructed a sundial for the latitude of Darlington, and this was erected by permission of the mayor, on his private lawn.

Only one unpleasant incident occurred on the encampment. Usually, the citizens of the towns where the cadets have camped have been most considerate and hospitable. While at Camp William H. Evans the people of Darlington showed the cadets every kindness, but one afternoon two young bloods of the town, who were rather contemptuous of this playing at soldiers decided to run over the sentinel on one of the posts and drove their spirited horse upon the bayonet of the cadet on guard. Fortunately, the animal was not seriously injured.

The baccalaureate sermon at Darlington was preached by the Rev. Albert S. Thomas, Class of 1892—now Bishop of the Episcopal Diocese of South Carolina—on June 22; and the Commencement exercises were held on Carolina Day, with an address by the Honorable R. B. Scarborough.

In December, 1901, the South Carolina Inter-State and West Indian Exposition, an ambitious and creditable undertaking for a city the size of Charleston, was opened, and continued until the following summer. It offered an excellent opportunity to advertise the Citadel and every advantage was taken of the occasion. A college exhibit to show the work and activities of the cadets was installed in the Palace of Agriculture, and two cadets were detailed each day to show this exhibit to visitors and give information about the Academy. Among the organizations that visited the exposition were the corps of cadets of the V.M.I. and the V.P.I., with both of whom the Citadel battalion appeared in joint parades and in sham battles. The encampment was omitted in this summer of 1902, due to the many interruptions caused by the Exposition, and this time was given to academic work. For the third successive year, a Citadel graduate was called on to preach the baccalaureate sermon. This year, the preacher was Rev. John G. Beckwith, a young man of talent and great promise, whose life was cut off a few years later (1907). The Commencement address was delivered on June 30, by Rev. J. A. B. Scherer, whose theme, "The Voice of Charleston," was treated with rare appreciation.

And thus endeth the Second Decade.

CHAPTER IX
THIRD DECADE: 1902-1912

THE decade from 1902 to 1912 in the history of the Citadel was marked by several beginnings and changes, and showed a continuous progress in its affairs up to a definite peak. The enrollment increased from 127 in October, 1902, to 236 for the session of 1911-1912, and the physical plant was surprisingly enlarged from cramped, ill-furnished quarters to a size which confessedly made it complete, so that, in 1912, the superintendent could assure the legislature in all good faith that "The Citadel would never need to ask for an appropriation for any more buildings." (How imperfectly did the superintendent see into the future!) But this interesting story will be told more in detail.

At the beginning of this decade, the Association of Graduates, always alert to the welfare of the Academy, began the publication of a quarterly *Bulletin* with the object of keeping the alumni informed about affairs at the Citadel and of enlisting their coöperation and support.

In the first issue of the *Bulletin,* January, 1903, was printed a resolution of the faculty, upon the advisability of which the opinion of the alumni was desired before being submitted to the Board of Visitors.

This resolution proposed a fundamental change in the curriculum of the institution by the introduction of electives in the senior year.

Two extracts from the article in the *Bulletin* will suffice:

"Owing to the fact that advances have been made in the work of the departments, requiring a greater amount of preparation or laboratory work, the course of studies, especially in the First (Senior) Class is too extensive for really good work in all departments. Even the best men in the class are unable to give the proper time to the study of all the subjects required, and it is safe to say that fifty per cent of this class can not do satisfactory work under the present schedule. They are pursuing courses in eight different departments, and are expected by the respective professors to do satisfactory work of course in every one of them. The time required in the classroom is twenty-seven hours per week, and when to this is added the amount of time consumed in military exercises, roll calls, etc., it becomes evident that no adequate amount of time can be given to preparation. . . .

"The elective system, by being limited to the First Class, would leave intact our three-year course as at present, thus not impairing the effectiveness of our curriculum in any vital particular; rather, indeed, rendering each department more effective by the proper classification of the students in their last year's work into groups where they can work to greater advantage, with more interest, with more time, and with greater concentration of purpose."

The resolution met the approval of the alumni, and was adopted by the Board. It went into effect with the graduating class of 1904, the nineteen members of which elected as follows: Engineering, 12; English, 4; Chemistry, 3. It was not until the year 1916 that the electives were extended to the Junior Class.

At the close of the academic session in June, 1903, the Corps was taken by rail to Rock Hill for the annual encampment and graduating exercises.

Camp Wade Hampton was pitched on the grounds of the High School (now the Winthrop Training School for Teachers). Besides the usual military work, Captain George H. McMaster, the newly appointed commandant, paid special attention to the rifle shooting, testing the merits of the Krag rifles recently issued to the cadets. A practice march of eight miles was taken to the concrete power dam on the Catawba River at the invitation of the engineer in charge, William S. Lee, Class of 1894, The Citadel. Mr. Lee, who is (in 1932) one of the foremost engineers of the country, with an international reputation as the constructor of some of the greatest hydro-electric power plants in the world, made some of his early reputation by his skill, energy, and resourcefulness on this project on Catawba River.

An excursion to Charlotte, North Carolina, was planned for the Cadets, not only as an entertainment, but also to bring the Citadel to the attention of the people of this enterprising city. A pleasing incident of this trip was a call of respect on the widow of Stonewall Jackson, who showed great pleasure in seeing the beloved gray.

The baccalaureate sermon was scheduled for Sunday morning, June 28, to be preached in the open air on the campus. A heavy downpour of rain, however, occurred at the hour for the service, and Rev. F. W. Gregg, being a Presbyterian and not a Baptist, consented to deferring the sermon to the evening under a roof. Mr. Gregg was a graduate of the Citadel, Class of 1894, and was the fourth consecutive alumnus to be called on for the baccalaureate sermon.

The Commencement address was delivered in the auditorium of Winthrop College, by invitation of the president, Dr. Johnson. The speaker was General Edward McCrady, distinguished soldier, lawyer, and historian. This was the second time that General McCrady had addressed the cadets on commencement day—the first being in 1887.

At the Commencement on July 27, 1887—the session was prolonged beyond midsummer in those days—General McCrady spoke to the graduates at the German Artillery Hall. The subject of his address was the obligation of the graduate of a State college to return a service to the State.

"You have been educated in an institution provided, and, in a great part, supported by the State. Education is a thing of value in itself—of material and pecuniary value. Higher education, especially, differentiates those to whom it is given from those who do not receive it. Now surely the State has no right to select you or any others for these inestimable benefits simply for your own individual advantage. It has no inherent right to bestow personal favors. It must act with a single eye and purpose to the general good—the good of all. Therefore, if it has educated you, it has educated you for the public service, not for your own personal benefit." The service which these graduates owed the State was, as he proceeded to explain, an intelligent and active interest in State politics.

General McCrady's second address, delivered to the Cadets in the auditorium at Winthrop on June 30, 1903, was a masterly presentation of South Carolina's military contribution in the War of the Revolution. It might well be printed and taught in the history classes of our public schools today. Certainly the military student of today should know this:

"To recapitulate then, one hundred and thirty-seven battles, actions and engagements between the British, Tories, and Indians on one hand, and the American Whigs on the other, took place in South Carolina during the Revolution, and of these one hundred and three were fought by South Carolinians alone; in twenty others South Carolinians took part with troops from other States; making in all one hundred and twenty-three battles in which South Carolinians fought within the borders of their State for the liberties of America."

This address was probably General McCrady's last public appearance, he dying the following November.

Of course, this Citadel Commencement at Winthrop was while the girls were home on vacation.

More than twenty years later, the president of the Citadel, while making a passing visit to Winthrop, spoke to the girls at a noon-hour chapel service in the auditorium, and piqued their interest by telling them that while they were quite familiar with the fact that many girls had graduated at Winthrop, possibly very few of them knew that once upon a time a class of boys had been graduated there—that in fact a well-known citizen of their city, A. E. Hutchison, was one of those boys. They enjoyed the joke when told of the eighteen cadets who graduated there on the last day of June, 1903. It may be said in passing that another Citadel class was to graduate at another woman's college at a subsequent date.

During the vacation of this summer, the Citadel lost one of its most valued and faithful employees, Sergeant James Condon, drummer at the Academy for eighteen years. He was a retired Union soldier, and at the time of his death was reputed to be the oldest enlisted man in the Army, having joined it as a boy musician about the year 1840.

His intimate association with the cadets at the guard room placed him in a position to exercise great influence over them, and while he was a friend to everyone, he was not slow to show his disapproval of any dereliction of duty on their part. A soldier to the core himself, he had an innate dislike for unsoldierly conduct, and his reprimands, given in his own peculiar way, had a wholesome effect on the more thoughtless and careless of the cadets.

He was buried in Magnolia, where a suitable monument was later erected to his memory as "a tribute to his faithfulness," the funds for that purpose being subscribed by the graduates and cadets.

At an annual meeting of the Association of Graduates on December 9, 1903, a group picture of the Old Guard of the Citadel was exhibited, showing photographs of all (or as nearly all as could be obtained) of the alumni living in this jubilee year, 1903.

Thirty-nine individual pictures made up the group, at the top of which was E. L. Heriot, Class of 1849, who was thought at the time to be the senior alumnus. But when the venerable and beloved Bishop P. F. Stevens arose at the banquet in the mess hall to respond to the toast, "Our Alma Mater," unknown to himself the mantle of "oldest living graduate" had fallen upon his shoulders, for, just a few days before, in far off California, the soul of Edgar LaRoche Heriot had taken its flight.

The Old Guard was passing, as was eloquently brought out by J. E. Peurifoy who paid a fine tribute to the ante-bellum alumni, who, "in fair weather and foul have nobly upheld the Citadel, giving by their example a high inspiration to the younger graduates who must now begin to carry the good work forward."

The secretary also noted this interesting incident of the meeting:

"In its sixty-one years' existence, the Citadel has had but eight Superintendents. At the beginning, this officer could not, of necessity, be an alumnus of the Academy; but early in its history, for obvious reasons, it was sought to place at the head of its affairs one who should not only be thoroughly familiar with its methods, aims, and needs, but also one imbued with its spirit and the love of a son for his Alma Mater.

"Since 1858, with the exception of Gen. George D. Johnston, who was Superintendent from 1885 to 1890, an alumnus of the Academy has been in command at the Citadel. This post was filled from 1859 to 1861 by Major P. F. Stevens, Class of 1849, who resigned in the latter year to go into the Confederate service as colonel of the Holcombe Legion. During the War, it was Major J. B. White, Class of 1849, who commanded the

Corps and led it into battle at Tulifinny. It was our (Association) President, Col. J. P. Thomas, Class of 1851, who was called to the superintendency of the Academy when it was reopened in 1882; and our present superintendent, Col. Asbury Coward, Class of 1854, has occupied the position for the past fourteen years. The four other superintendents, three of whom were at the head of the institution during the first fifteen years of its history, were not Citadel men.

"It was a remarkable circumstance, not noted at the time, that at the annual supper the four Citadel Superintendents were seated side by side at the table.

"The members of the Association were fortunate to hear from each of them on the occasion."

During the year 1904, there were several events worthy of record. First, in January of that year, the legislature appropriated $10,000 for the installation of a modern system of hot-water heating in the Citadel buildings, and the wiring of these buildings for electric lights, thus getting rid of the serious menace of open fireplaces and kerosene lamps. These improvements were completed by the opening of the fall session.

On June 15, the Corps had a notable trip to the Louisiana Purchase Exposition at St. Louis, spending a week in specially provided barracks inside the grounds, and giving daily drills and dress parades for the public on the Plaza of St. Louis.

The cadets paid their own railroad fare, but their maintenance was provided by the Academy in lieu of the expense of the annual encampment. The cadets arrived in St. Louis early on the morning of June 17, and went immediately to the Exposition grounds for breakfast. Arrangements were made with one of the large restaurants in the grounds for meals for the cadets and the manager on this first morning had been quite anxious to please. He said, "Knowing your boys were from South Carolina, you will notice that I have had rice prepared for them for their breakfast!" It would have been too cruel to tell him that South Carolinians always had rice for *dinner*—the midday meal—but never for breakfast.

It goes without saying that the educational value of the stay of the cadets at the Exposition was very great. And it was gratifying to Colonel Coward to receive from Honorable D. R. Francis, president of the Exposition, a letter in which he said: "The soldierly bearing and uniform courtesy of the South Carolina Cadets during their stay with us will long be a pleasant memory with all our officials."

The Corps returned to Charleston on June 28 and on the 29th the baccalaureate sermon was preached at Grace Church by Rev. W. P. Witsell, Citadel alumnus of the Class of 1894.

The aptness of calling an alumnus to address the cadets on occasions of this kind was in Colonel Coward's mind when he introduced the custom in 1900. Mr. Witsell referred to this in his opening words:

"My young friends and brothers, just ten years ago I wore your uniform and was as you are now; and so I think that I can enter into and understand your thoughts and feelings tonight. I share your joy of accomplishment, and I can appreciate your regret at leaving friends and severing the heart-cherished associations of the past four years. And as you think of the future, while perhaps your minds may be somewhat bewildered by a sense of your unknown fortunes, I feel sure that in your hearts there unceasingly burns the bright light of inspiring and sustaining hope.

"And allow me to believe also that in your souls there is deeply implanted the true and manly desire to be somebody, and to do something worthy of your trained, educated, and potentially splendid manhood—that in the strife of life and on the world's broad field of battle you desire to be men and heroes, and not as dumb, driven cattle, or cowardly, shiftless men, or nonproducing drones. Therefore, fellow students and fellow men, I am here not to attempt to pronounce an oration, but to point out to you, if perchance God has given me ability to do so, some fundamental truth which may serve as a lantern to your feet in the dark and slippery places in your pathway through life."

The Commencement exercises were held next day, Honorable George S. Legare being the speaker.

The trip of the cadets to St. Louis was undoubtedly an effective piece of advertising, and helped make the Citadel more widely known.

Another excursion into (literally) foreign fields was made during the summer vacation. This was a trip by Colonel J. Colton Lynes, of the Citadel faculty, a versatile linguist, to Cuba and Porto Rico, with a view to getting a number of Spanish-American students to enroll as cadets.

Eight young Cubans, who had previously received high school training in English, were admitted to the Fourth (Freshman) Class as a special section, and in the following year three others were entered.

It may be said here that only two or three of these boys showed any inclination or ability to acquire the training and education given at the Citadel, and the experiment was abandoned.

The session of 1904-1905 opened in October with an enrollment of 149 which practically filled the cadet barracks to their capacity. At the beginning of the session, Captain McMaster was relieved at his own request as commandant, and was succeeded by Captain William H. Simons, U.S.A., who was a Citadel graduate of the Class of 1890, and was peculiarly well fitted, therefore, for the position.

In the spring of 1905, an unexpected piece of good fortune came to the Citadel in a dividend from bonds of the Charleston Exposition which had been assigned to the Citadel by about thirty merchants and citizens of Charleston. The amount realized—$2,200—sufficed to construct and equip a gymnasium in the space between the main building and the West Wing, which was named in honor of the chairman of the Board of Visitors, one of the contributors.

The encampment and Commencement of 1905 were held in Columbia, June 15-30. The baccalaureate sermon was preached in Washington Street Methodist Church by the pastor, Rev. A. N. Brunson, Class of 1888, and the Commencement exercises were held in the opera house, Dr. W. Spencer Currell being the speaker. Dr. Currell, who later became president of the University of South Carolina, was then professor of English at Washington and Lee University.

The Citadel was now clearly on the upgrade.

The enrollment was increasing from year to year, and the legislature was more kindly disposed towards it.

A matter of great concern to the institution now began to take definite form. This was the return to the State of the central police station, which had been constructed on the King Street end of the Citadel plot just after the earthquake in 1886. If the Citadel were to grow, this property would be required for its use.

The session which opened in the fall of 1905 had an enrollment of 168 cadets, the largest number since the first corps in 1882.

In January, 1906, the city council of Charleston offered to return the site and donate half the cost of the guard house to the State for the use of the Citadel. The legislature accepted the offer, and appropriated $22,000 for the purpose, to be paid in three installments.

Thus the first measure for the enlargement of the Academy since the War of Secession was adopted.

Upon the recommendation of Captain Simons, a practice march in the spring of 1906 was substituted for the summer encampment. Following is his official report of the experiment:

"The itinerary of the march was as follows:
April 26, Charleston to Ten Mile; distance .... 10.1 miles.
April 27, Ten Mile to Strawberry Station; distance ... 13.6 miles.
April 28, Strawberry to Pinopolis; distance .... 11.1 miles.
April 30, Pinopolis to Carn's Cross Roads; distance ... 12.4 miles.
May 1, Carn's Cross Roads to Summerville; distance .. 7.4 miles.
May 2, at Summerville. Day devoted to field exercises.
May 3, Summerville to Ten Mile; distance .... 13.6 miles.

May 4, Ten Mile to Charleston; distance .... 10.1 miles.

Total distance marched .... 78.3 miles.

"As will be noticed, the marches were all short, and nothing was carried by the Cadets except the rifle and equipment, canteen and haversack, with meat can, tin cup, knife, fork, and spoon. Ample camp equipage was obtained from the State, and good transportation was provided.

"When the Cadets first heard of the march, there was apparently a decided sentiment against it; but when it was actually commenced, there were only two cadets who, through pernicious home influence, shirked the entire march.

"The conduct of those Cadets who participated in the march was excellent, and considerable grit was displayed by the entire command. One cadet fell out the first day, and it was necessary to transport him on a wagon the remainder of the trip. His ailment, however, was contracted before the march. Two Cadets were taken sick at Pinopolis, and were transported from that place to Summerville. Only two other cadets fell out of the column while on the march; and this was between Pinopolis and Carn's, after they had marched half the distance, and on a particularly hot day.

"While every effort was made to make the march as easy as possible—guard duty being very light, the command marching light, and practically no fatigue duty except making and breaking camp—I consider that the Corps has made a remarkable record for itself, for the ages of the Cadets must be considered, and also the fact that they were taken directly from their books and that it was an entirely new experience to nearly all of them.[1]

"In the matter of instruction, I am of the opinion that the Cadets derived considerably more benefit from the march than they would get from the four summer camps of a cadetship. The improvement of the Corps in making and breaking camp and the manner in which the Cadets looked out for their health and comfort was noticeable from day to day, and I be-

---

[1] An amusing incident may be recorded to show that they retained, at least, their sense of humor. The second day's march from Ten Mile to Strawberry was through a sparsely settled country. The day was hot, and the boys were footsore from the first day's hike. Along the country road, many curious negro boys came out of their cabins to see the strange sight of soldiers on the march, and it was a frequent question from the cadets: "Say, boy, how far is it to Strawberry?" The answers were indicative of the African idea of distances. One might reply, dubiously, "It's about five miles," and a few miles further on the next would say, "Oh, it's about six miles." The trudging soldier boys soon gave up hope of getting information, but the questions continued mechanically. After a while one darkey made the encouraging answer, "It's jest about three miles," so that a mile further on there was a faint inflection of hopefulness in the question addressed to the next black boy, "Say, you boy, how far is it to Strawberry?" "It's jest three miles, mister.' Whereupon, one perspiring wag in the ranks called out in mock resignation, "Thank God, boys, we are holding our own!"

lieve that any Cadet who made this march would be a valuable addition to any National Guard company or any body of newly organized troops taking the field."

The pernicious home influence, referred to by Captain Simons, was not a new experience for the commandant. As a cadet, he had been familiar with the methods employed by the shirkers in the school. Always there were some parents who, while lauding the "splendid discipline at the Citadel," wanted it abated in the case of their own particular Johnnie.

They were not all like the parent of one cadet (he is now a prominent business man, and has served as president of the Association of Graduates) who tells this joke on himself: While at home on leave, when he was a cadet, he was complaining of the many punishment tours that he had to walk, for offenses that he thought were quite inoffensive, and after amplifying his tale of woe with great effect, as he thought, he begged his father to write for an honorable discharge for him. His father listened to the story without comment, but seemingly with sympathy, and merely said he would let him know next day. The boy returned to barracks at retreat, and next morning paid several visits to the guard-room anxiously expecting to get the letter requesting the president to give him a discharge. At last it came—a strange box with it. Hastily opening the note, he read: "Dear Son: I am sending you three pairs of shoes. These will last you through your punishment tours."

Early in his official connection with the Citadel as military officer, Captain Simons undertook the important but laborious task of revising the basic regulations of the Academy, separating the general authority of administration from the special—the code of Academic Regulations, containing the powers and functions of the Board of Visitors for the regulation and control of the institution, and, distinct from this, the Blue Book, which contained the regulations for the interior police and discipline of the cadets.

Not only this. Finding that the records of the Citadel had never been systematically compiled, he undertook to perform this really stupendous task. This extra work, which Captain Simons performed as a pure act of service for his Alma Mater, and without any thought of remuneration, literally took hundreds of hours of his own time. But as a result of his labors, the Citadel has a card-index file which contains, it is believed, a record of every cadet who has ever matriculated at the Academy.

A notable event of the year 1907 was the visit of the corps of cadets to the Jamestown Exposition, near Norfolk, Virginia. The cadets went into camp on the grounds, and gave daily military exercises, but had ample time to study the Exposition and to make excursions to some of the numerous historic places in the vicinity.

The baccalaureate sermon was preached by Bishop A. M. Randolph, in old St. Paul's Church, Norfolk, on June 23, and the Commencement exercises were held in the auditorium at the Exposition on June 27, Governor Ansel, of South Carolina, delivering the address.

In the spring of 1908, Captain Simons conducted the Corps upon its second practice march, leaving Charleston on April 1 and returning on the 14th. The nights en route were spent at Clementia Springs, Adams Run, Horse Shoe Mines, and a camp of five days held at Walterboro. Setting out on their return, the cadets made camp at night at Cottageville, Givhans, Summerville (two nights), and Ten Mile.

The health of the Corps on the march was excellent, and everywhere the cadets were received with the greatest hospitality.

Just before the battalion set out on April 1 for the sixteen-mile hike to Clementia, one of the cadets received a telegram calling him home, and did not rejoin the Corps until its return to the Citadel. In order that this cadet should not altogether miss the benefits of the march, the commandant calculated the number of hours the battalion had spent on the march and assigned him that much of his leave time in walking around the quadrangle. The cadets followed his progress from day to day with much interest, and would call from the gallery asking him if he had "reached Adams Run yet?"—or cheered him when he got to Cottageville and Givhans. This boy may have had to go home on a legitimate errand, but there is no doubt that he wished he could have gone on the march.

Two Citadel alumni were called back to their Alma Mater for the Commencement exercises of 1908—Rev. George H. Atkinson, for the baccalaureate sermon, and Professor William E. Mikell, dean of the law school of the University of Pennsylvania, for the annual address. The beginning of the sermon was rendered a little too solemn by an eclipse of the sun—not total, but of enough digits to diminish very appreciably the daylight, so that lamps had to be lit in the church.

The end of the session was made memorable by the retirement of Colonel Coward from the superintendency. To quote the *Bulletin:*

"A change in the headship of the institution is an event which at any time is not only of great importance to the Academy and of concern to the alumni, but likewise of considerable interest to the public. This is particularly so in the present instance on account of Col. Coward's prominence and the fact that his retirement will close the longest administration in the history of the Citadel, and one which has been marked by the consistent growth of the Academy in efficiency and influence, and in the confidence of the people. That a large measure of this success has been due to the wise, thorough, and kindly administration of Col. Coward, the alumni of the Academy know and appreciate.

"Himself an ante-bellum graduate of the Citadel, he brought with him at the beginning the love of a son for his Alma Mater, and in the long years of his superintendency, in which it has been his one thought and concern, he has become identified with it. It will be difficult to think of the Citadel without Col. Coward, and when he leaves he will carry with him the hearty and affectionate good wishes of all the hundreds of young men who have been under him in the eighteen years since October 1, 1890."

At the final dress parade, the chairman of the Board of Visitors presented to Colonel Coward a loving cup as a token of the Board's appreciation of his long and faithful stewardship.

The detail of Captain Simons was due to expire in October, and this occasion was taken to make some public acknowledgment of his services. This testimonial was in the form of an officer's saber in a beautifully chased scabbard, presented on behalf of the Board by Colonel Coward.

The Board of Visitors elected Major O. J. Bond, Class of 1886, to succeed Colonel Coward as superintendent. Major Bond had been elected as an assistant professor just after his graduation in 1886, and had been continuously a member of the faculty since that time, so that the affairs of the Academy were familiar to him. In the first issue of the *Bulletin* after the opening of the session in the fall, he gave the following account of the completion of the King Street Extension—the former guard-house which the city of Charleston had turned back to the State for the Citadel's use—and the opportunity which the Academy now had to develop:

"November 1, 1908.

"The opening at the Citadel was marked by a number of changes which will be of interest to the alumni. While the Academy is in essentials the same with which the graduates and ex-cadets are familiar, in some respects it may be considered as having entered upon a new era in its history. Considered merely as a pile of buildings, the addition of about forty per cent to its capacity by the acquisition of the remodeled old City Station House, makes it a Greater Citadel.

"This extension has also given opportunity for large improvements and increased cadet barracks in the main building, to which—for disciplinary reasons—the dormitories of cadets must always be limited.

"The removal of the hospital and the library from their former inadequate space in the barracks to new, well-furnished, and commodious quarters is a great improvement. The alumni, especially, may take pride in the well-equipped library and the large, comfortable reading-room, which were furnished by their contributions to the building fund. Certainly no more substantial or important improvement was made, and it

Colonel Oliver James Bond

will be a pleasing memorial to their active interest in the prosperity and success of their Alma Mater.

"The session opened with the largest attendance in its history, there being now 193 cadets enrolled. This number could easily have been increased beyond 200, but the faculty wisely rejected the applicants who could not pass the entrance examinations—about 10%—believing that it is better to go slowly on a sound basis than to encourage or permit an unhealthy inflation. The experience of the past five years has been that 40% of the freshman class fail to advance, and the aim of the Citadel will be not to increase its numbers too rapidly at the sacrifice of the quality of its recruits, but the building up of a limited corps of fine material.

"There have been a number of changes in the departments. Besides a new superintendent, there is also a new commandant, three new professors, and two new assistants—the latter being graduates of the class last June, a new policy.

"Arrangements with the Chief of Ordnance, War Department, have just been completed whereby new equipment of 200 rifles and accoutrements and two 3.2-inch field pieces with equipment will be issued to the Academy in the next few days.

"Some few changes in the curriculum have been put into effect this year, and others will no doubt occasionally be made to keep the Citadel abreast of educational times, but it will be the general aim to develop it along the well-marked line of its progress up to the present time."

Early in his administration, the new superintendent took up a question which he considered of supreme importance. In the issue of the *Bulletin* for February 1, 1909, he brought the matter to the attention of the alumni thus:

"When our alumni are asked the question: 'Where did you graduate?' the great majority answer, 'At the Citadel'—and not 'the South Carolina Military Academy.' Similarly, most Army men, we believe, who are alumni of the United States Military Academy call themselves graduates of West Point.

"These titles have the merit of being short, and have been so long used that it may even be said that they are more widely known than the longer official titles. But may there not be also a more or less acknowledged prejudice in the minds of the alumni of these institutions against an implied inferiority in the name to the words 'college,' 'seminary,' 'institute,' or even 'school'?

"Since the national military academies were founded, a multitude of preparatory military schools which go under the name of 'Military Academies' have come into existence, and, indeed, the multiplication, popularity, and excellence of these 'Academies' has since become so great that the name is now practically identified with secondary schools. Among the 451 institutions of higher learning in the country, the two Government schools at West Point and Annapolis, and the Citadel are the only three which today retain the name 'Academy' officially in their title, and all three of these, it seems, prefer to use the better known names by which we have called them rather than their official names.

"The institutions at West Point and Annapolis are sufficiently distinctive as Government schools for the technical training of Army and Navy officers to give a special significance to the word 'Academy.' But can such be said of any other military institutions in the country? These exist, not to supply the needs of the military and naval establishments of the Government, but, in their highest ambition, to give young men 'a complete and generous education—that which fits a man to perform justly, skillfully, and magnanimously all the offices of a citizen, both private and public, of peace and war.'

"So far as the writer knows, the Citadel has the distinction of being the only one of these military institutions with college curricula which is known officially as an 'Academy.'

"The Pennsylvania Military College—one of the eight distinguished military schools in the class with the Citadel—seems to be an example in which the name 'Academy' was recognized as a handicap or misnomer, and hence dropped. It was known from its foundation in 1862 until 1892 as the Pennsylvania Military Academy. In a historical sketch of the institution published in its catalog, the following explanation of the change of name appears:

" 'Changes, improvements, and additions have marked the passing of recent years, and the present equipment of buildings offers excellent opportunity for the collegiate education and military instruction of a large corps of cadets.

" 'In order that the name of the institution might properly indicate that the Legislature had invested the Board of Trustees with collegiate powers and privileges, the Court of Common Pleas of Delaware County, Pennsylvania, Dec. 12, 1892, changed the corporate title to Pennsylvania Military College.'

"That the word 'academy,' with its now universally accepted significance is a handicap and a misnomer at the Citadel is the general opinion of the officers of the institution. Book publishers with whom the Academy has dealt for years still send sample copies of elementary text-books in arithmetic, English, and languages to the professors for use in their classes. Parents write to matriculate twelve-year old sons. The impression of the usual intelligent and inquiring visitor to Charleston is that we prepare students to enter West Point. Even prominent educators in our own State have expressed surprise and sometimes incredulity when informed of the grade and character of the work done here. But it is especially when the Citadel wishes to extend its clientage beyond the borders of the State that the most serious handicap is felt. In a magazine where twenty 'academies' advertise, it does not occur to the general reader that the South Carolina Military Academy is in a class by itself—and such a statement in our advertisements would not be considered a model of modesty, although these are the exact words of one of the United States Inspector-Generals. Unless it happens to be known, its name ranks it among the preparatory schools.

"The original Act of the Legislature establishing educational institutions at the Citadel and Arsenal was passed Dec. 20, 1842, and was entitled:

" 'An Act to Convert the Arsenal at Columbia and the Citadel and Magazine in and near Charleston into Military Schools.'

"This Act did not specifically give these institutions any names. But the Arsenal and Citadel being well-known military posts, it was implied that the schools established there would take the names of these posts.

" 'The Citadel,' therefore, was the official as well as the familiar name of the present Academy from its foundation in 1842 until it was closed at the end of the War in 1865. At the beginning of the War of Secession, an Act was passed by the Legislature, the first section of which read: 'That the Arsenal Academy and the Citadel Academy shall retain the same distinctive titles, but they shall together constitute and be entitled "The South Carolina Military Academy." ' When the Citadel was reopened in 1882, there was no expectation of also reopening the Arsenal, and hence no real reason for retaining the partnership name, so to speak. At any rate, all through the years of its post-bellum operation, the institution has worn its long name of 'South Carolina Military Academy' only as a top-coat—within, it is the same old beloved Citadel.

"Is there any weighty reason why the name with which it is identified should not be restored as its official title? The legal questions involved should not be formidable, since the property of the Academy is not held in its own name, but is the property of the State.

"As no one mind sees all sides of a question, however, and 'in the multitude of counsel there is wisdom,' the *Bulletin* would like an expression of opinion from the members of the Association, and cordially invites all who have any interest in the matter to correspond with us. We would like to see the title made officially 'The Citadel,' with a sub-title, 'The Military College of South Carolina.' How do the graduates and ex-cadets, ante-bellum and post-bellum, feel about it?"

The proposition aroused a good deal of interest, and at the meeting of the Association of Graduates on June 30 was discussed at length. It was finally decided to make a further statement to the alumni through the *Bulletin,* and send a reply card for an expression of opinion on the question to every member of the Association.

Colonel Bond presented the matter in more extended form in an article in the August *Bulletin,* and in November announced that of the antebellum alumni eleven voted for The Citadel, and two for South Carolina Military Academy. Of the twenty-two classes since 1886 which expressed a choice, all cast a majority vote for The Citadel, a minority vote for the longer title being cast in eight classes.

In view of the preponderating preference for the retention of the name The Citadel, the president of the Association addressed a communication to the Board of Visitors reciting the main points of the discussion and giving the results above, and requested the Board to take whatever action it thought proper in presenting the matter to the legislature.

It may be stated here that the Board approved the change of title, and the matter was brought up in the legislature at its session in January, 1910, with the result that an act was passed fixing the title as

The Citadel
THE MILITARY COLLEGE OF SOUTH CAROLINA

At the end of the session in June, 1909, the Corps went into camp at Fort Moultrie on Sullivan's Island with the object of giving the cadets instruction in the heavy artillery of the coast defense batteries. Under the direction of Major Landers and the other officers at the fort, the cadets were taught how to operate and fire the 10-inch disappearing guns and the 12-inch mortars.

The Commencement exercises began in camp on June 27, when the baccalaureate sermon was preached, by Rev. B. G. Murphy, Class of 1896. This service was unusual in being held in the open, under the shade of the poplar trees surrounding the Army Y.M.C.A. building.

The graduating exercises were held in the pavilion on the Isle of Palms, the speaker being Colonel Coward, whose address was unique in one respect. It is not likely that another instance can be cited of an address being delivered a second time after an interval of half a century, the treatment of the subject being equally applicable and forceful at the later date.

This was true of Colonel Coward's address, which was delivered first before the graduates on April 11, 1860. From the *Bulletin,* November 1, 1909:

"The attendance at The Citadel at the beginning of the session again broke all records. Two hundred and forty-four cadets and recruits were enrolled for duty and matriculation. Of the latter, fourteen were rejected on account of insufficient preparation.

"The corps is organized as a battalion of five companies of forty-five men each, a cadet band of twenty-one pieces, and the staff.

"The large increase in attendance made necessary the provision for additional dormitory rooms in barracks. The Commandant, Lieut. W. St. J. Jervey, gave up one of the three rooms in his private quarters, the armory was turned into barracks, and three instructors changed rooms from the Main Building to the Extension, thus adding five rooms to cadet quarters.

"The Superintendent and the Commandant are now the only officers living in the Main Building. The West Wing is occupied by four families and in the Extension are seven other officers, the matron of the Mess Hall, and the matron of the Hospital. With every square foot of space occupied, the Institution has reached the limit of its present capacity.

"With the growth in the size of the corps, not only must additional quarters be provided for the cadets, but also accommodations for more officers, and new classrooms.

"The first can only be obtained by the addition of a fourth story to the Main Building. The latter can best be provided for by the erection of a new building on Meeting Street which shall be the counterpart of the King Street Extension, and which will give room for other urgent needs.

"An architect is at present making plans for these needed improvements, and the Board of Visitors will probably consider this matter at their meeting in December, and make a report to the next Legislature."

The organization of a cadet band was an important development in the military department.

A complete set of band instruments was obtained on loan from the State, and Chief Musician Ensey, First Band, Artillery Corps, at Fort Moultrie, was engaged to instruct the cadets. Sergeant Ensey had an excellent reputation as a band director, and the First Band was reputed to be one of the best in the service. His work with the cadets was most satisfactory, and the cadet band became an institution in the school which it could ill afford to do without.

In January, 1910, the Board of Visitors extended an invitation to the General Assembly, then in session, to visit The Citadel with a view of showing the legislators the work of the institution and its need for the fourth story and the Meeting Street extension.

The invitation was accepted, and on January 26, a large delegation of the lawmakers and their wives and daughters was entertained by The

Citadel and the city. A repast was served them in the mess hall, the souvenirs being cards with a picture of The Citadel as it would be when the fourth story was added and the Meeting Street extension built.

On May 1, the *Bulletin* had this announcement:

"At a special meeting of the Board of Visitors during 'Battleship Week,' the contract for building the fourth story of The Citadel was let to Mr. J. T. Snelson, of Charleston, for $26,883.

"Mr. Snelson expects to begin work early this month and to have his contract completed by Sept. 15th.

"The Legislature did not feel justified in making appropriation this year for the building on Meeting Street, but next session it is hoped that this will be provided for.

"At the invitation of the city of Greenwood, the annual encampment in the summer of 1910 was held in that progressive upcountry town, June 1 to 15.

"The location in a grove of oaks behind the Presbyterian Church and not far from the business center of the city, was very convenient, and the city authorities had made every arrangement for the comfort of the cadets, including water, baths, latrines, electric lights in the tents, and even telephones at headquarters and at the mess hall.

"On the arrival of the Corps, a sumptuous repast was served by the ladies of Greenwood, which was not only fully appreciated by the boys but also by the quartermaster's department, as time was thus obtained to get our own service into condition for its important duties.

"The military work of the encampment consisted of the usual camp guard instruction, daily morning drills and practice marches, daily target practice at the range, daily dress guard-mounting and dress-parade, a sham battle with problems in attack and defense of position, and at the last, an exhibition drill by the companies and battalion, and a review of the Corps by the Board of Visitors. The cadet band added very greatly to the success of the encampment, and besides their daily duties in connection with the guard-mounting and dress-parade gave a concert every morning at camp headquarters.

"The target shooting was made a special feature of the military work. Instruction in gallery practice had been previously given at The Citadel and all the members of the three lower classes had been given a course on the rifle range of the 3d Regiment near Charleston, but at Greenwood a range had been constructed to continue this important feature of the work."

The Commencement exercises began on June 12, with the baccalaureate sermon by Dr. John O. Willson, president of Lander College. Dr. Will-

THE CITADEL.
The Military College of South Carolina.

SOUVENIR.
Visit of the General Assembly.
Jan. 24, 1910.

son is a former cadet, being one of the famous cadet company which left The Citadel in 1862 and served so brilliantly in Virginia.

The graduating exercises were held on the evening of June 14, in Waller Hall, at Lander College. The annual address was made by Captain Ellison A. Smyth, of Greenville, another old Citadel boy—or rather Arsenal cadet. Captain Smyth gave a most instructive and interesting account of the cadets in the last year of the War.

A most pleasing incident of the encampment was a characteristic gift from Dr. John O. Willson, which is explained in his letter below:

"Greenwood, S. C., June 14, 1910.
"Col. O. J. Bond, Superintendent.

"Dear Sir: Since 1905, I have had a provision in my will for a small fund to furnish yearly a neat 'ring for that member of the senior class of the Citadel Academy who shall be adjudged by the ballot of the entire Corps of Cadets to be the manliest, purest, and most courteous of his class—this ballot to be conducted under rules prescribed by the faculty of said Academy, consulting with the Board of Visitors. And if in the judgment of the said Faculty and Board the ballot be in danger of abuse, then award shall be made by such means as may be fixed by said Faculty and Board.'

"If this meets your approval, and the approval of any besides who should decide the matter, I should be glad to begin the award of the ring at once —for the class now graduating. . . . And I would be pleased to consult as to the design of the ring, or ask you to determine that.

"Sincerely yours,
"JOHN O. WILLSON."

Dr. Willson's gift was accepted, and in the years to come will be not only an incentive to the cadets to practice the high soldierly virtues which he names, but will be a happy memorial of an old cadet who illustrated them throughout his life.

Such is the origin of the John O. Willson Ring, the presentation of which has been a feature of each annual commencement since the Greenwood encampment. The twenty-two wearers of this ring at the present date form a roll of alumni of which The Citadel is particularly proud.

The ring is of gold, with a bloodstone setting, and inscribed with the three words, Manliness, Purity, Courtesy.

The year 1911 marks The Citadel's attainment of one of the peaks of its history. The *Bulletin* of February 1 contains this article:

"The year 1911, it is hoped, marks the completion of The Citadel. The plans worked out some years ago have been carried forward steadily, and there remains only the Meeting Street Extension to finish the design and complete the equipment of the institution so far as buildings are concerned.

"In the future there will be no thought of further expansion, for the limits of The Citadel are definite. Within these bounds, however, it hopes to realize a large ambition. The graduates of just a few years ago would hardly recognize the Alma Mater of their day in The Citadel of 1911. Even to one who has lived in it through all its years of vicissitude, its development seems to partake of the marvelous.

"When the first class after the War graduated in 1886, there were only the Main Building and the East Wing. At that time there were doubts of the continued existence of the school; and that the West Wing would ever be rebuilt was only a dream of the most sanguine. In 1887, the State gave to the city of Charleston the lot on the King Street end of The Citadel property on which to construct the new police station. When this was built, there was a gap between it and the main building where the ruins of the old burnt wing stood as a sad memorial of the 'Yankee occupation.' Yet the very next year, a Yankee Congress voted the State $77,250 to restore the burnt wing and equip the institution for its work as a college. Then the Police Station became an undesirable companion, and in 1908 it was bought by the State at half its value from the City, and was remodeled for the use of the Military College. In 1910, in order to accommodate the growing size of the Corps, the Legislature provided for the fourth story of the Main Building, and this year the Meeting Street Extension will go up to provide for the last urgent needs of the college, and to complete the beautiful architectural design. When this is built The Citadel will present an appearance so imposing that its superior from an architectural point of view will be hard to find, and within this compact plant, the ambitious task will be undertaken of developing a military college of which the State may well be proud."

At the session of the legislature in January, an appropriation of $50,000 was made for the Meeting Street extension, and work was begun upon the building at once.

At the same session, an act was passed conferring authority upon the Board of Visitors to grant the degree of civil engineer to graduates of The Citadel. This subject had been discussed with the alumni through the medium of the *Bulletin* for a year or more, and in view of the special emphasis which had been laid upon the course in civil engineering since the foundation of the school, and of the eminence which many of the alumni had attained in the profession, the propriety and desirability of The Citadel's granting the degree was universally recognized. A number of alumni of national reputation, members of the American Society of Engineers, expressed themselves as desirous of getting the degree from their Alma Mater.

In 1908, the Society for the Promotion of Engineering Education—of which The Citadel is a member—appointed a committee to consider and report on the practicability of simplifying the character of degrees conferred in engineering. In the report of this committee, which collected information from ninety-two institutions teaching engineering, it recommended that the four-year college course in engineering should lead to the degree of bachelor of science, and that "the professional engineering degrees, C.E., M.E., etc., should be given only to graduates who present satisfactory evidence of professional work of superior quality extending over not less than three years, and who submit a satisfactory thesis." It also recommended "that the same degrees may be given to engineers as honorary degrees; but great care should be exercised in awarding them."

The policy recommended by the association was adopted by The Citadel faculty.

In the year 1911, a change in the policy of the encampments was also made, the cadets being taken for two weeks of practice in rifle shooting on the range of the National Guard about five miles above the city, April 1-15. The camp was named in honor of Mayor R. G. Rhett. The results of the rifle practice were so satisfactory that the next spring the corps was taken again to the range for two weeks of the same instruction. Camp Thomas, however, after nine days of work, was brought to a tragic close by the death of one of the cadets. On the afternoon of April 9, 1912, the program called for rapid fire at 300 yards. The superintendent and commandant were in the butts observing the operation of the targets by the cadets. The protection afforded by the earth bank was ample, but in some strange way, one of the bullets was deflected downward by striking a piece of the woodwork overhead, and instantly killed Cadet Louis Dotterer, who was operating one of the targets. This deplorable accident terminated the rifle practice at the range.

*   *   *   *   *

In these early years of the twentieth century, the rehabilitation of the South was proceeding slowly.

In the first decade of the century, educational advance, which depended immediately upon economic progress, seemed almost like Mark Twain's account of the movement of a certain Swiss glacier, which his guide book told him was travelling downward into the valley below. Why not take a seat on the glacier and return to his inn in this way, rather than by the rugged pathway that he had laboriously climbed to the top? So he sat—until he read on further in his guide book that he would reach his destination in about 167 years.

In South Carolina, the educational system was top-heavy—there being an overdevelopment of institutions of higher education compared with the common and high schools. The result was that the colleges were doing some of the work that should be done by the preparatory schools.

In the report of the superintendent of The Citadel in 1911, he says:

"The standard of scholarship at The Citadel continues to be maintained, and the requirements for admission are as strict as the condition of our common and high school system seems to warrant. While admission to The Citadel is not based upon a system of units, it has been found from an inspection of the certificate submitted by the freshmen of this year that the average number of units offered is about ten. The difficulty which has been experienced at The Citadel with candidates for admission is that they are not thoroughly prepared in the elementary branches. A considerable number of candidates every year show a lack of proficiency in arithmetic, elementary algebra, English grammar and composition, history of the United States, and descriptive geography. Students coming from some of the best high schools and who have advanced much beyond these requirements are frequently as deficient as others in these essential branches. This lack of preparation will account largely for the great number of failures which occur in our freshman class. Last year, out of 115 cadets in this class, only fifty-three made the required average at the end of the session. The casualties in the higher classes were not so large, only one senior and three juniors failing. In the sophomore class, however, the casualties were also heavy—twenty-eight out of seventy, or 40%."

The long road upward of raising the standard was to be painfully followed during the next two decades; and now, in this year 1932, after some seemingly encouraging advances, we wonder whether we have really progressed very much, or, like Mark Twain's glacier, gone a little way downward.

For the Commencement exercises this year, 1911, two of The Citadel's alumni were called back for the baccalaureate sermon and the annual address—Rev. T. M. Hunter, of Baton Rouge, Louisiana, and Dr. James P. Kinard, of Winthrop College.

The opening of the session of 1911-1912 was advanced from October 1 to September 20, in order to permit closing ten days earlier in June. Up to this time the school had opened at the beginning of October, and the closing exercises had taken place in the heat of midsummer—indeed, in the 1880's in the latter part of July. (The class of 1886 graduated on July 28th.) The change to an early June graduation was altogether wise.

The graduating exercises in the summer of 1912 began, therefore, on June 9, with the baccalaureate sermon by Dr. John Kershaw, rector of St. Michael's Church. Dr. Kershaw was an ex-cadet of Civil War days. The graduating exercises were held on June 13, Dr. Arthur W. Goodspeed, of the University of Pennsylvania, making the address.

An interesting feature of this commencement was the conferring of the first new Citadel engineering degrees upon five of its alumni, who had qualified to receive it. They were:

R. M. Walker, Class of 1886, engineer and contractor, Atlanta, Ga. Thesis: "The Determination of the Bearing Power of Soils."

W. A. Leland, Class of 1886, specialist in concrete and dam construction, Rockingham, N. C.

W. M. Smith, Class of 1889, civil engineer and designer of dams, New York. Thesis: "Design of Masonry Dams."

P. B. Bird, Class of 1890, U. S. Engineer, Jacksonville, Fla. Thesis: "Examination and Survey of St. John's River, Fla."

L. L. Gaillard, Class of 1890, civil and electrical engineer, New York.

W. S. Lee, Class of 1894, civil and electrical engineer, Southern Power Co. Thesis: "Parallel Operation of Hydro-Electric Plants."

F. G. Eason, Class of 1906, U. S. Drainage Engineer for South Carolina. Thesis: "Report of Preliminary Drainage Examination of Great Pee Dee River Valley."

And so this narrative comes to the end of the Third Decade of the new Citadel's history, which can be fitly closed with the closing of the career of one of her most loyal and distinguished sons:

Colonel John Peyre Thomas
Born March 17, 1833—Died Feb. 11, 1912
First-honor graduate of The Citadel, Class of 1851
Professor at The Citadel, War-time Superintendent of the Arsenal
Founder of the Carolina Military Institute, Charlotte, N. C., 1873
Superintendent of The Citadel, 1882-1885
Member of Board of Visitors
Historian of his Alma Mater

"If I perchance were called upon to name
One man I knew who stood four-square alway,
Who held his honor dearer than his fame,
John Peyre Thomas is the name I'd say.

What more? Could any language add
A title to a character so clean,
That held as odious all he knew was bad,
And through all, ill or wrong, kept right, serene?"

W. H. GIBBES.

CHAPTER X

FOURTH DECADE: 1912-1922

IN these reminiscences of The Citadel, a great deal has been said about the military features of the institution. It would be entirely erroneous, however, to think of The Citadel as having for its primary purpose the turning out of soldiers. In 1877, Colonel Thomas published a *Sketch of the South Carolina Military Academy,* in which he set forth in a single phrase the purpose of the institution as he conceived it: "A complete and generous education—that which fits a man to perform justly, skillfully, and magnanimously all the offices of a citizen, both public and private, of peace and war." These words were adopted by The Citadel as the foreword on the front page of its catalog.

It may be said, in passing, that Colonel Thomas, who was usually meticulous in his quotations, did not in this case give Milton's exact words, which were as follows:

"I call, therefore, a complete and generous education that which fits a man to perform justly, skillfully, and magnanimously all offices both public and private, of peace or war."

Colonel Thomas's interpolation of the word "citizen" we would not wish to alter, as it is needed to emphasize the fact that The Citadel has for its primary object the education of her sons for citizenship. But even from a literary standpoint, one whose appreciation of letters was strongly influenced by Colonel Thomas's teaching may be pardoned for preferring his well-balanced phrase.

Notwithstanding a notion that may exist in some quarters—and sometimes exploited, perhaps, to the disparagement of The Citadel's educational scheme, the military training at the institution is primarily a means to an end. The college which has lax disciplinary features, which does not exercise an active supervision over the living conditions of its students, and has small responsibility for their conduct, is relieved of its most onerous cares and anxieties.

But at The Citadel, the idea prevails that all of education does not consist of recitations, study hours, and laboratory periods. These may be ninety per cent, more or less; but The Citadel scheme definitely takes into consideration the inculcation of certain personal habits and traits which are as important in life as purely mental training. The military system is believed to be peculiarly well adapted for this training.

Undoubtedly, there are persons who object—conscientiously, perhaps— to any training which "interferes with the development of the student's individuality." Especially are they opposed to "standardizing the college output." Others, who, in spite of their own youthful errors, have turned out well, wish their sons to "sow their wild oats, and learn to stand on their own feet as soon as possible."

For graduate students in the universities and professional schools, no one would recommend placing any restrictions upon their movements or time. College boys from sixteen to twenty years of age, however, are still in their formative period—not yet masters of their wills or habits.

As a rule, the men who have been through the mill of military discipline at college believe in it.

From an article by Lieutenant Colonel A. M. Hitch, president of the Association of Military Colleges and Schools of the United States, in *Cosmopolitan* for September, 1932:

"A few months ago the Office of Education at Washington, under the direction of Secretary Wilbur, published a pamphlet (No. 28) that contains a remarkable study on this subject. Over ten thousand questionnaires were received from R.O.T.C. graduates of fifty-four institutions in forty states. These men graduated from 1920 to 1930; accordingly most of them had been out of school long enough to have a perspective.

"Of the answers 97.1% said that the R.O.T.C. military course of study had a definite educational value of its own; 94.9% considered its contribution unique and its training justified by the results. With convincing unanimity the ex-cadets insisted:

(1) that military training developed in them confidence, coördination of mind and muscle, and contributed practical training in leadership;

(2) that disciplinary exercises taught courtesy and respect for authority;

(3) that training contributed to an interest in national affairs, and to a conviction that there are duties and obligations of citizens in times of peace;

(4) that it gave an appreciation of the importance of health, neat appearance, and erect carriage;

(5) that it taught in theory and practice the principles of coöperation, organization, and the art of command; developed self-control, dependability, and orderly habits;

(6) that (and in this 93.6% concurred) it was conducive to a sane patriotism and an enlightened attitude toward peace, a working familiarity with modern military weapons acting as a sobering offset to any romantic conception of warfare."

It is said that in industrial plants one of the problems of the manager is how to dispose of the by-products. Sometimes it is possible to sell the waste, so that it will really pay the dividends. In some cases, it has even happened that the by-product has afterwards turned out to be the chief product, as has been told, for instance, of kerosene and gasoline. Before the days of the automobile, when the kerosene lamp was the only shining light, it became desirable to remove the highly explosive products from kerosene in order to make it safe for domestic use. Gasoline was one of the undesirable by-products for which the manager must find a use. In the popular mind, in that primitive age, cleaning clothes was about the only use it could be put to; but the clothes did not get soiled fast enough to use up the supply. When, however, the ingenuity of man devised an engine in which gasoline served as an ideal fuel, and also found better illuminating contrivances than kerosene lamps, the process of manufacture was reversed. Gasoline, from being the vexatious vixen in the refinery, became the undisputed mistress of the establishment, and the elder sister, Kerosene, sank to the lowly estate of the poor relation.

All of which is suggestive, perhaps, of the military institutions of the country which had been going along for so many years doing their work of training young men for citizenship—incidentally teaching them something of the art of war—and suddenly an insane fanatic fires a torch at Sarajevo which directly turns the world topsy-turvy and eventually sets all our colleges in the United States to turning out soldiers instead of scholars. A remarkable story!

In its annual military session—in camp or on the march—The Citadel had been doing for many years a work which the War Department was beginning to recognize.

Due to the proximity of The Citadel to the coast defenses on Sullivan's Island, the feasibility of instructing cadets in the operation of the heavy ordnance was apparent. For the second time, therefore, the corps was taken to Fort Moultrie in the summer of 1913 for practice with the 10-inch rifles and 12-inch mortars. Next year, the time was given to a practice march of about eighty miles, Orangeburg being the objective. And in the spring of 1915, the entire two-weeks' work was devoted to rifle practice, so that a cadet who graduated this year had had instruction in rifle shooting for two fortnight encampments, a practice march of eighty miles, and one fortnight in instruction in coast artillery work.

In 1915, Captain S. J. Bayard Schindel, War Department inspector, said in his report:

"The Cadet Battalion of four companies and band—strength, 207—was inspected at Camp William W. Lewis, Mt. Pleasant, S. C., not far from Sullivan's Island.

"The camp was nicely located in a grove, and well drained. The cadets were fed in practically the same way as in barracks. Rifle practice (range firing) was being carried on in the National Guard range about a mile from the camp. Practice carried on up to include 1,000 yards—the course for the organized militia being fired. Interest and results satisfactory. . . .

"The establishment of annual camps with army officers detailed as instructors is believed to be an ideal solution of this important problem of providing officers for our volunteer companies in time of war. They should be carried on in every school, and every facility permitted under the law should be accorded by the War Department."

The next year, Captain Schindel was again the inspector, and said:

"The cadet battalion was in camp at Mt. Pleasant near a very good target range. Target practice was going on. . . . The camp was inspected and found to be in an excellent state of sanitation and police.

"Extended order drill was seen, and improvement noticed in the use of signals. The bayonet exercise was also improved.

"Two outpost problems and the development of a hostile position were all satisfactory. The field engineering work at camp was excellent. This particular instruction has only lately been developed.

"The institution has been most persistent in the endeavor to improve military instruction. Under the wise and careful administration of the Superintendent, coöperation with the War Department has been greatly increased. He has pride in the institution and his efforts are now bearing fruit.

"It is recommended that in view of the probable number of vacancies in the regular army, that not less than ten of the graduating class, such as are selected by both the Superintendent and the Commandant, acting conjointly, be appointed second lieutenants under the provisions of the Bill recommended and proposed by the Committee on Military Affairs, U. S. Senate."

In the spring of 1917, the annual inspection was made by First Lieutenant Frank K. Ross, who reported in part as follows:

"The following exercises and ceremonies were witnessed, all excellent: Guard mount, battalion inspection, battalion parade, battalion drill, calisthenics, and bayonet drill.

"A field problem consisting of an advance guard developing an attack of position, and a rear guard. This exercise was handled entirely by cadet officers, and was executed with vigor and in a highly satisfactory manner, demonstrating a well grounded knowledge of minor tactics.

"Ten members of the senior class have been designated for commissions as second lieutenants of the army, and it is recommended that the number be increased to fifteen, as the graduates are especially fitted. The institution seems to be working in complete harmony with all things military, and the authorities have instilled a high degree of spirit amongst the cadets. This college is undoubtedly one of the distinguished military institutions of the country."

The National Defense Act of June 3, 1916, was of special interest to The Citadel. It offered the opportunity to a number of recent graduates to enter the Regular Army, or to enroll in the Officers Reserve Corps. The formation of the Reserve Officers Training Corps—the R.O.T.C.—in the colleges of the country made a special appeal to The Citadel which had for so many years been doing work along the lines now adopted by the War Department for the training of reserve officers in the colleges.

There was much criticism of the National Defense Act from the pacifist element of the country, who talked of Prussianizing the United States by compulsory military service. Even in South Carolina, which has never been noted for its pacifistic tendencies, the superintendent of The Citadel had more than once to explain to the people of the State that the institution was primarily a State educational institution, and not a training school for officers of the Army. Writing in *The State*, August 23d, he said:

"At present, there is an unusual, if not undue, amount of talk about compulsory military training—which is not the same thing as compulsory military service. Not two years ago, President Wilson stated that the young men of America would not give up the best years of their life to military service. If, however, while pursuing their course in civil education they can at the same time receive efficient military training, the matter is easily settled. At The Citadel, we have long believed that not only is there no waste of time due to military instruction, but also the punctuality, prescribed study hours, regularity and industry promoted by the military system effect an economy of time, and in addition form a valuable part of the student's education for all walks of life."

In the summer of 1916, upon the invitation of Captain Wm. C. Harllee, of the U. S. Marine Corps, the cadet rifle team was sent to the Marine Corps range at Winthrop, Maryland, for special instruction in rifle shooting preparatory to entering the National Rifle Match which was held that year in October, at Florida's giant range at Black Point on the St. John's River. The funds to defray the travelling expenses of the team to Winthrop were donated by the Association of Graduates, and Captain Harllee —himself an old Citadel cadet—who is greatly interested in rifle shooting, promised that the team would receive the best of coaching. The result at Black Point amply justified the trip; for although the cadet team did not rank very high relatively—for it was shooting in the fastest company in the world—its score was altogether creditable. Fifty-five teams competed, the Marines winning first place, with a percentage of 84.86. The Citadel

Dress Parade at Old Citadel

was in thirty-first place with a score of 78.86. The last team had an average of 53.75.

In January, 1917, the General Assembly, while in session in Columbia, was invited to visit Charleston and The Citadel, and see from personal inspection something of the State's chief seaport city and her military college.

The occasion was most happy for both guests and hosts. Representative Hoyt, Speaker of the House, said he had only one criticism to make. This was:

"The Citadel is too modest. Members who inspected the barracks were surprised that the institution is not asking for an appropriation to thoroughly overhaul the entire building."

The criticism was confessedly deserved. The institution had been reared in the hard school of poverty. It had learned to endure hardships, and had not had the opportunity of acquiring expensive habits. It had received its meager appropriations with thanks, and had not complained loudly enough, perhaps. The Ways and Means Committee had always asked if the requests for appropriations were the least that the institution could exist on, and had expected an affirmative reply.

On March 4, 1917, President Wilson, who had been reëlected "because he kept us out of war," began his second term, and very soon was to show himself the most militant of all the presidents. He called for military force without limit—force to the uttermost. Certainly, Wilson was not temperamentally a pacifist, and when the issue was joined he clearly recognized that if the world was to be made safe for anything, it would be by the iron hand of might.

On April 8, Congress promptly followed his leadership and declared a state of war to exist between the United States and Germany.

Immediately upon the declaration of war, The Citadel was ready to give all its energies to the government in preparation for the conflict. On April 7, the following telegram was sent to Major General Leonard Wood, commanding the Department of the East, who was coming to Charleston to take charge of the newly-formed Department of the Southeast:

"Gen. Leonard Wood, Governor's Island, New York.

"I respectfully offer to you, with the approval of the Board of Visitors and the Governor of the State, all the military facilities of The Citadel, The Military College of South Carolina, including instructors and cadets, for such uses as you may desire to make of them in training recruits for service.
(Signed) O. J. BOND,
Colonel and Superintendent."

The people of the State, true to its traditions, strongly supported the President.

Governor Richard I. Manning set a notable example by giving his five sons to the service of the country—one of whom never returned from France.

One interesting incident illustrative of the patriotic ardor of the people may be recorded in connection with the Summer School for Teachers at Winthrop College, which was an annual event of considerable importance. More than a thousand teachers of the State—most of them women—customarily spent a six-weeks' vacation at this educational session in professional and cultural training. In June, 1917, the chief topic of discussion on the campus was, of course, the war.

Dr. Johnson had drafted the superintendent of The Citadel for a course in mathematics, but his position as head of the State military college gave him a wholly undeserved prestige as a military instructor, and he was amazed to find that a large number of the women teachers of the State wanted more to learn to drill than to solve problems in mathematics.

Concealing his ignorance, Colonel Bond sent to The Citadel for Lieutenant Garey's new Plattsburg Manual, commissioned Dr. Kinard of Winthrop faculty—an old Citadel man—and Miss Nancy Campbell, the capable director of the music department, as lieutenants, and proceeded to teach a company of 125 young women, of draft age, first, the school of the soldier, and then the manual of arms (with wooden guns). The matter of uniforms was easily settled; it was an army in white.

The first drill period was before breakfast—in the cool of the morning, when there were few spectators; the other in the late afternoon when the company showed off before the admiring gaze of the public. Before the Summer School ended, this company of women could execute so creditably the evolutions of a company and the manual of arms that Mrs. Manning, the governor's wife, was invited to review them.

Nor did these women find this training child's play. When asked why they were willing to undergo this unnecessary drudgery in vacation-time, they gave this reasonable answer: they were teachers; there were many boys in their schools; the country was at war, and the boys were on the battlefield; how did it feel to be a soldier? Being true teachers, they wanted some knowledge about these things founded in their own experience.

The class which graduated at The Citadel in the summer of this year numbered thirty-three members. The records show that every one of them entered the military service. Six received commissions in the Regular Army; thirteen others were commissioned in the Marine Corps, which was a very popular service with The Citadel cadets about this time. In this connection, it is worthy of record that in these years of uneasiness about our military unpreparedness so many Citadel men received commissions in the

Marine Corps that jealous Congressmen from other states had made the reputed partiality shown this Southern school the subject of investigation by a Congressional committee.

The session of 1917-1918 opened on September 20 with an enrollment of 259 cadets. No modifications of the scholastic work were found to be necessary, but Lieutenant Garey's high-pressure work in the military department was continued by his successor, Colonel Ralph R. Stogsdall, U. S.A., retired.

December 20, 1917, marked the seventy-fifth year of the founding of The Citadel. In spite of the fact that so many of the alumni were engaged in fighting for their country, it was decided that those left at home would celebrate in a modest way the Diamond Jubilee.

The exercises were held in the chapel of The Citadel on the evening of that day, speakers having been selected to tell of the three phases of Citadel history, the past, present, and future.

Colonel Coward was the first speaker, and no one more capable or appropriate could have been selected. Just before Dr. John Kershaw was to speak on the present, a large service flag was unfurled by Mrs. James H. Holmes and Mrs. T. F. McGarey, mothers of two graduates of The Citadel then on the battlefield in France. The flag displayed in large figures, formed of stars, the number 210—that being the number of alumni at that time in the military service.

In speaking of The Citadel of the future, Mr. Lesesne, president of the Association of Graduates, said: "There is a fruitful field of endeavor before our graduates in helping to build up the Greater Citadel of tomorrow" —using words prophetic of an undertaking of undreamed-of consequence to be carried through by the alumni a few years later.

At the close of the exercises, the superintendent read a telegram from Dr. John O. Willson: "Alma Mater—congratulations! May your thousandth birthday find you sending out such sons as you are today."

★ ★ ★ ★ ★

The year 1918 did not open auspiciously for the Allies. Relieved of pressure on the eastern front by Russia's revolution, Germany had brought many fresh divisions to the Western Front, and was so confident of a victorious drive to Paris that even the date of the movement was advertised rather than concealed.

The United States was now in the war; but in spite of the assuring words of a great and lovable citizen, a million men could not spring to arms overnight. There had to be long, persistent, and systematic preparation. Our War Department set about putting an American Army into France as soon as possible, but the most optimistic could not hope for a victorious decision before 1919. The early contingents of American troops were going to fight with the English and French divisions, and in the first contingent there were Citadel men. Some had gone over and joined the Allies before the United States became involved in the struggle. In 1915, Lieutenant Montague Nicholls—well remembered at The Citadel—fell in battle while serving with the British Royal Artillery in Flanders. The first officer from South Carolina to give up his life on the battlefield after the entry of the United States into the war, March 1, 1918, was Lieutenant John H. David, Citadel, Class of 1914.

But we will let one of the most gallant of The Citadel officers in the American Army tell something of his fellow alumni in the service. Speaking at the Greater Citadel banquet in Columbia, given by the Association of Graduates on the evening of January 21, 1920, Colonel B. R. Legge said:

"In responding to the toast, 'The Citadel Man in the Service,' my thoughts go back to that historic morning of Jan. 9, 1861, when a Citadel man, carrying out the orders of a Citadel man, pulled the lanyard that sent the first shell of the war of secession screaming to its mark, the U. S. Steamer *Star of the West*.

"It is salient. When the layman finds that fact on history's page the first question that comes to his mind is: 'Is this only an incident that this detachment of cadets manning a battery of 24-pounders under command of their superintendent, should stand out as the unit that initiated the war of secession, or are they destined to play a greater part?' Curiosity prompts him to turn the pages—he finds that of the 240 graduates, 200 were officers in the Confederate service, and that forty-three gave their lives on the field of honor. He finds a story of leadership and heroism and unflinching service through four years of bitter fighting, which culminated on May 9, 1865, when Capt. J. P. Thomas, a Citadel graduate commanding a detachment of cadets near Williamston, S. C., fired the last shot of the war by any body of organized troops east of the Mississippi.

"Does he wonder that when the call to duty came in April of 1917 that the sons of that institution came to the colors just as they did in 1861, just as they did in 1898, and just as they will always do when the call of country comes?

"It is interesting to note that history has repeated itself.

"When the first American convoy sailed on June 13, 1917, there were a number of Citadel men with it.

"They were with the artillery brigade that pulled its gun up through the mud of Lorraine, and going into position near Bathlemont, sent America's first shot into the German lines.

"On March 1, 1918, the first American success in arms was gained

THE STORY OF THE CITADEL

when the First American division was holding the lines north of Toul. A small crisis came—it was met. I will read you the order it called forth from the lieutenant-general commanding the Thirty-second French corps:

'Ansauville Sector, Headquarters,
March 2, 1918.

'First Army, Third Corps,
'General Orders No. 119.

'On the first day of March at daybreak, the enemy pulverized the front line trenches and dugouts occupied by the Eighteenth American regiment with a heavy fire of Minnewerfers and 210's. They then attacked in six columns under the protection of a rolling barrage. All instructions which had been given were faithfully carried out. The Americans withdrew to the edge of the zone under fire, then delivered a strong counter-attack.

'The Boche realized the force of the American blow; he retreated to his position leaving fifteen dead, including two officers and four prisoners.

'The troops of the Thirty-second Army Corps, proud to be fighting by the side of the generous sons of the great republic who have hastened to support France, and with her to save the freedom of the world, will understand by this example of superb courage and coolness the full meaning of the promises made by the entry into the conflict of their new brothers in arms.

'The general commanding the Thirty-second Army Corps heartily congratulates the first American Division, and in particular the Third Battalion of the Eighteenth Infantry, as well as the American artillery whose precise and opportune action contributed to the success.

(Signed) 'PASSAGA,
Lieut.-Gen. Comdg. 32 A.C.'

"When the Third Battalion of the Eighteenth Infantry counterattacked in the grey mist and smoke of bursting shell that morning, First Lieutenant John H. David, class of 1914, was first out of the trenches and first to strike the Boche. He fell at the head of his platoon, on the field of honor, 'a gallant gentleman.'

"From that morning until November 11, Citadel men were in every active phase of America's participation.

"They were with the units that stemmed the tide at Château Thierry and Montdidier.

"They were at Cantigny.

"They were at the hinges of the great counter-offensive on July 18.

"They were at Juvigny and Fismes and on the Chemin des Dames, and with the assaulting units when the St. Mihiel salient fell.

"They were fighting it out on their ground in the bitter struggle in the Argonne Forest, and bridging the Meuse on November 8.

"They were with the first American division that made that bold dash under cover of night across the face of twenty kilometers of the enemy's positions and formed up at dawn in the closing hours of the greatest war in history on the heights of the historic Sedan.

"Three hundred and fifteen in the service of their country; 126 in the Expeditionary Forces; six killed, seventeen wounded—slackers none.

"The war is over. Citadel men still serve, from the Island of Mindanao to the Steppes of Siberia.

"The mills of the old institution grind slowly—the product changes not. It stands for the same principles, the same ideals—solid citizenship, unquestioning loyalty, unflinching service."

✧ ✧ ✧ ✧ ✧

At the commencement on May 24, The Citadel welcomed back as the speaker for the graduating exercises Major William Cain, professor of mathematics at the University of North Carolina. The Citadel had given Professor Cain his military title, and at the University he had retained it through all the years from 1889, when he left The Citadel to enter upon the larger field at Chapel Hill.

The graduating class numbered twenty-eight men, and, as in the preceding year, every one of them went into the military service. Thus the two war classes—1917 and 1918—bore out the tradition as told by Colonel Alfred Aldrich, a member of the famous cadet company of the 60's. When the Charleston Light Dragoons returned from the Mexican border in February, 1917, the battalion of Citadel cadets was at the head of the escort of honor. Colonel Aldrich was with a group of Northern tourists on the veranda of the Charleston Hotel, and witnessed the procession. He told them a humorous incident of a former parade, when a visitor asked a Citadel cadet how many of the graduates of the school had served in the Confederate Armies, and received the matter-of-fact reply, "Only one hundred per cent."

In the summer of 1918, a great war-time training camp for college men was held at Plattsburg, New York. Very properly, the units from the various colleges were broken up and distributed among the various companies. The time was not propitious for institutional rivalries. Whatever loss was entailed by the lack of college spirit was more than made up by the lessons of individual responsibility and of working for a common cause. No cadet from The Citadel or the Virginia Military Institute could gain any prestige from the reputation of his college, but only by his personal merit. There was an abundant appeal to the spirit of rivalry—but it was to one's company; and Citadel, Lehigh, Boston Tech, and Virginia Military Institute men worked valiantly together against their college mates in other organizations.

While at Plattsburg, an inspiring message came from far-off France to The Citadel boys. It was a cablegram from three captains in the 26th Infantry, 1st Division, Barnwell R. Legge, James H. Holmes, and Julius A. Mood, all of whom had been cadet officers in the battalion of Citadel Cadets.

"France, June 17, 1918.
"South Carolina Corps of Cadets,
"Plattsburg, N. Y.
"We are depending on the standard you will set in a new state.
(Signed) "LEGGE, HOLMES, MOOD."

Just one month later, after celebrating a holiday on Bastille Day in Paris, the First Division was rushed up the line near Soissons, where Foch launched his great drive against the German flank, July 18-21.

The Twenty-sixth Infantry was in the hottest of the fight. Holmes fell on the 19th—leading his company in its fourth assault against the stubbornly yielding enemy. Mood fell on the 21st. Legge was the senior officer in command of the regiment on the last day, all his superior officers having been killed or wounded.

At Plattsburg, in June, 1918, there was also a notable gathering of college presidents. They had a very vital problem to discuss; their students being of draft age, would they have any college classes at all when the opening day of the fall session arrived? Would not all their students be in the army training camps, being prepared to fight their country's battles?

The solution of their problem was radical but simple. The colleges themselves would be the training camps. The college professors would be retained and the students would continue some of their regular courses, but experienced Army officers would take charge of their instruction in military science and tactics, and as soon as they were ready for active service in the field, they would be sent to join the colors. The plan adopted was known as the Student's Army Training Corps.

When the session of 1918-1919 opened, therefore, an unprecedented ceremony took place in Charleston. The student bodies of The Citadel and the College of Charleston assembled jointly on the campus of the latter, and took the oath to serve their country faithfully as soldiers.

The Citadel barracks were crowded to their capacity. Three hundred and fifty young men were enrolled for the session.

Under the National Defense Act of June 3, 1916, the cadets at The Citadel, practically all of whom were enrolled in the R.O.T.C., obligated themselves to take military training as a part of their college course, but their responsibility to the discipline and the regulations of the college was not disturbed. It seemed desirable to The Citadel that this status with the government should be retained, and efforts were made to have it continue to operate as an R.O.T.C. unit under the Defense Act. The War Department, however, thought it undesirable to make exceptions.

Under the S.A.T.C. (Student's Army Training Corps), the students were enlisted in the military forces of the government, and were directly under the control of the Army officers detailed for their instruction. College regulations were superseded by the Articles of War. Presumably a cadet breaking barracks could be confined in the guard-house on bread and water, or, if he deserted, he might be tried by a court-martial and sent to Leavenworth.

The regular class work naturally suffered, as the students expected to be sent off to join the troops as soon as they qualified for such duty. Indeed, it requires little imagination to foretell the baleful effects on scholarship if the experiment had been prolonged. Fortunately, the colleges had hardly had time to organize their S.A.T.C. units before gratifying reports from the Western Front presaged the early dissolution of the German power, with the happy result that the war psychology was only an evanescent phase of college experience.

There was great rejoicing in the early hours before dawn on the morning of November 11 when an immense concourse of people assembled on Marion Square to celebrate the signing of the Armistice and the end of the world's greatest tragedy.

The number of Citadel graduates in service during the World War was 316. The roll of ex-cadets is incomplete, but was probably as great. It may be of interest to note the rank of The Citadel graduates in the service. The highest rank was that of colonel, of which there were eight. There were nine lieutenant colonels, twenty-three majors, one commander, and five lieutenant commanders (Navy), ninety-eight captains, sixty first lieutenants, sixty-three second lieutenants, nine naval lieutenants, ten sergeants, seven corporals, and twenty-two privates—the last being the most noteworthy item in the list.

THE CITADEL ROLL OF HONOR IN THE WORLD WAR
DIED IN THE SERVICE

First Lieutenant John Hodges David, Jr., Class of 1914, 18th Infantry, 1st Division. Killed in battle near Ansauville, France, March 1, 1918.

First Lieutenant Arthur Thomas Elmore, Class of 1917, U.S.M.C. Mortally wounded at Belleau Wood, June 24, 1918; died July 13.

Formation in Front of Old Sally-port

First Lieutenant George Hampton Yarborough, Jr., Class of 1916, U.S. M.C. Mortally wounded at Belleau Wood, June 24, 1918; died June 27. Distinguished Service Cross and Navy Cross (posthumously).

Captain James Hill Holmes, Jr., Class of 1915. 26th Infantry, 1st Division. Killed in battle near Soissons, July 19, 1918. Distinguished Service Cross (posthumously).

Captain Julius Andrew Mood, Jr., Class of 1916, 26th Infantry, 1st Division. Killed in battle near Soissons, July 21, 1918. Distinguished Service Cross (posthumously).

Lieutenant Colonel Robert H. Willis, Class of 1908, Air Corps, 7th Army. Died in service, September 13, 1918.

WOUNDED IN THE SERVICE

Private D. H. Laird, Class of 1916, 6th Infantry. Severely wounded September 11.

Second Lieutenant B. A. Sullivan, Class of 1911, 118th Infantry, 30th Division. Wounded August 14, in Belgium.

First Lieutenant James Anderson, Class of 1916, 118th Infantry, 30th Division. Wounded October 11.

First Lieutenant J. C. Cogswell, Class of 1917, U.S.M.C. Severely wounded June 8. Distinguished Service Cross, Navy Cross.

First Lieutenant D. A. Holladay, Class of 1917, U.S.M.C. Wounded at Belleau Wood, June 24. Wounded near Soissons, July 19.

First Lieutenant C. T. Smith, Class of 1911, 114th Machine-Gun Battalion, 30th Division. Wounded October 17.

Captain E. B. Hope, Class of 1917, U.S.M.C. Wounded at Château Thierry, June 6. Croix de Guerre. Distinguished Service Cross.

Captain R. W. Hudgens, Class of 1915, 118th Infantry, 30th Division. Severely wounded.

Captain H. Hutchison, Class of 1915, 118th Infantry, 30th Division. Wounded slightly.

Captain B. R. Legge, Class of 1911. 26th Infantry, 1st Division. Chevalier of the Legion of Honor after Aisne-Marne Offensive. Also the Croix de Guerre. Wounded (slightly) October 4, Meuse-Argonne Offensive. Distinguished Service Cross, Croix de Guerre with palm.

DECORATED

In addition to those mentioned above as winning awards, the following men were decorated:

First Lieutenant H. E. Losse, Class of 1913, 322d Infantry, 81st Division. Croix de Guerre.

First Lieutenant A. M. Parrott, Class of 1913, 371st Infantry. Croix de Guerre.

Captain R. C. Hilton, Class of 1915. Machine-Gun Company, 9th Infantry. Croix de Guerre.

Captain G. C. McCelvey, Class of 1911, 47th Infantry. Distinguished Service Cross.

Captain J. F. Moriarty, Class of 1917, 5th Regiment, U.S.M.C. Croix de Guerre, Fourragère.

Captain E. P. Norwood, Class of 1917, 6th Machine-Gun Battalion, U.S.M.C. Twice awarded Croix de Guerre and Fourragère.

There were many citations for gallant conduct on the field of battle given to other Citadel men which are on record.

THE BUILDING OF THE GREATER CITADEL

And now we come to the most remarkable period in the history of the institution, the building of the Greater Citadel, an accomplishment, in the short space of a dozen years, of an expansion and growth undreamed of by the most sanguine of its friends.

In this great project there were many decisions involved, all of them essential to the realization of the final objective.

In the first place, a definite limit to The Citadel at Marion Square had been reached. In its beginning in 1830, it was only a great two-story, quadrangular fortress. In 1850, a third story was added to give more barrack rooms for cadets; then the two wings were built in 1854—and these sufficed for more than fifty years. It was in 1908—far along in the history of the reorganized Academy—that the city of Charleston turned over the central police station to be reconditioned for the use of the institution. Then followed, in a few years, the addition of the fourth story to the main building for more cadet rooms, and finally, the Meeting Street extension. In 1912, therefore, The Citadel, at Marion Square, had reached the end of its development—to accommodate a cadet corps limited to 325 students, apportioned among the several classes as follows: sixty-two seniors, sixty-eight juniors, eighty sophomores, and one hundred and fifteen freshmen. This would allow for a loss of thirty per cent in the first-year men, fifteen per cent in the next year, and a ten per cent loss in the junior year. In some respects, this may have been ideal; it would certainly have been select. But in fifteen years all the enrollment figures above were more than doubled.

Another thing: At Marion Square The Citadel had no campus or drill ground except a small part of the square, which was constantly used by the public as a cross-thoroughfare. The intersection of King and Calhoun

THE STORY OF THE CITADEL 193

Streets was already the center of the business section of the city, and was becoming each year more congested and unsuitable for the location of an institution of learning. As in the experience of some other colleges, it was getting time for The Citadel to move.

But if a move were to be made, a wide field would open for the selection of a new site. Doubtless many cities would compete in offers to locate the college in their limits or vicinity. It was recalled that early in the history of the United States, Congress came very near locating the National Military Academy within the borders of South Carolina—at Great Falls, in Chester County. No doubt, Chester would be glad now to have the West Point of the South established on the banks of the Catawba River.

But the upcountry already had more than its share of the State's educational institutions, and The Citadel was too much identified with Charleston in associations and tradition to be wantonly torn from its mother city.

A suggestion had been made by Governor Pickens during the War of Secession to move the Academy to Sullivan's Island—but that was not with the purpose of giving it larger opportunities. Just across the bay, one sees from the battery the inviting shores of James Island, with beautiful sites and unlimited field for expansion. It is, however, remote, except by water transportation.

But there happened to be an adequate and ideal location, on a site of unsurpassed beauty—not distant, nor even in the suburbs, but well within the limits of the city—easily accessible to all its conveniences, but yet enough apart from its business bustle and distractions for the establishment of a seat of learning.

If one will examine the fifty-year-old map of Charleston, prepared in 1883 for Mayor Courtenay's Centennial Address, he will find the one-mile track of the Washington race course occupying the larger part of the land between Moultrie and Grove Streets now occupied by Hampton Park. It was in this locality that the famous duels took place in ante-bellum days. The last remnants of the race-course buildings were removed for the erection of the Exposition Palaces in 1900, and afterwards the city acquired the property for a public recreation ground—Hampton Park.

Just west of the race course was the Revolutionary estate of John Gibbes, afterwards known as the Rhett Farm, and acquired by the city council from the Charleston Library Society. It was the purpose to develop this tract of seventy-six acres as an extension of Hampton Park, and the eminent landscape architect, Olmsted, of Boston, laid out the avenues circling the property, passing over Indian Hill and winding under the stately old live oaks. For some years previous, this terrain had been more or less familiar ground to the cadets, whose field exercises under the instruction of the

Army officer had been carried out here on a number of Saturday morning drills.

Now, in the fall of 1918—when The Citadel was overcrowded with its great influx of S.A.T.C. students—a committee of the Board of Visitors, consisting of Mr. J. P. Thomas, Colonel J. G. Padgett, and Colonel James H. Hammond, assisted by Mr. D. G. Dwight, president of the Association of Graduates, and a number of local alumni, approached Mayor T. T. Hyde and the city council on the proposition of offering to the State the above tract for the building of a Greater Citadel.

Evidently the time was ripe; for such a resolution was passed by the council—and the first great step had been taken.

The first public announcement of the city's offer was made in the columns of the *News and Courier* on December 11, 1918, by Mr. T. P. Lesesne, city editor of that paper, and past president of the Association of Graduates. Both in its news columns, and editorially, this paper now set out to promote favorable public opinion for the acceptance by the legislature of the city's offer.

A bill for that purpose was introduced into the legislature which convened in January, 1919.

To the Citadel men in the General Assembly, assisted by many influential friends—among whom one of the most valuable was the State senator from Charleston County, Mr. Arthur R. Young—large credit must be given for the overwhelming sentiment in that body in favor of the bill. It passed the House by a vote of ninety-eight to seven, and the Senate by a vote of twenty-seven to seven.

The act made an appropriation of $300,000 for the first building. It also provided that the Board of Visitors might sell the buildings and site at Marion Square within the next three years and devote the proceeds to further construction at Hampton Park.

And now the responsibility for the accomplishment of the great project rested upon the Board. On April 17, 1919, the Board met at The Citadel and did two important things—engaged attorneys to arrange the transfer of title to the new site from the city to the State, and appointed a building committee to look into the best method of proceeding with the construction.

The building committee, consisting of Mr. Thomas, Colonel Padgett, and Colonel Hammond, decided "to go slow and make no mistakes." Calling into consultation Mr. R. M. Walker, of Atlanta, an alumnus of The Citadel (1886) and an engineer and contractor of experience and responsibility, they met at The Citadel on May 1 and went over the situation. After careful consideration this important decision was made:

"The Committee would recommend to the Board that a responsible firm of engineers be engaged to take complete charge of the proposition—this firm to employ, first of all, a competent landscape architect to make a study of the site in connection with the plans for present and future development, to make drawings and specifications for the several buildings as funds become available for their construction, to advise with the Board as to the engineering problems involved in the whole scheme, and as to the awarding of contracts, and to supervise and pass on the work of construction. In this way, it is hoped that nothing will be done in an ill-advised way, but that there will be a coördination of plans looking to a final harmonious development of the whole project."

The Board approved the committee's report on May 7, and began an investigation into the qualifications of a number of engineering firms.

Two months later, it was announced that the firm of Lockwood, Greene & Co., of Atlanta, had been engaged to take charge of the construction of the Greater Citadel.

On November 7, the contract for the new buildings was awarded to Mr. R. M. Walker, of Atlanta, Georgia, who submitted the lowest of six bids received.

Mr. Walker was the first post-bellum graduate of The Citadel—being the first-honor graduate of the Class of 1886. When he heard of the Greater Citadel project, he became greatly interested in it, and when he was selected by the Board to be its builder, he intended that the construction of the new and "more stately mansions" for his Alma Mater should be the crowning work of his life. In his own words, he "wished to make it his monument." But just at the beginning of the great program of building, he was stricken with influenza, and died at his home in Atlanta on February 7, 1920.

On the editorial page of the Atlanta *Journal* appeared this estimate of him:

"In the death of Mr. Robert Murdoch Walker, Atlanta loses a citizen of outstanding ability and devotion. Since casting his lot with this community some twenty-five years ago, he has given continual proof of his interest in its highest welfare. A builder in the broadest sense of that noble word, he wrought his ideas into the structure of things civic as well as into the great edifices of steel and stone. His work, whether for the public or for private interests, was ever of that substantial and distinctive quality which characterized the man himself. Never slighting the minutest detail and never failing in the largest grasp of principle and design, he has left an enduring monument in every building entrusted to his construction and a cherished memory in every heart that had the good fortune of his friendship."

The Charleston Engineering and Contracting Company succeeded to the contract thus untimely terminated by the death of Mr. Walker, and in March Mr. Thomas, chairman of the building committee, announced that the construction of the buildings at Hampton Park would go forward with as much expedition as possible.

The *Citadel News* of February, 1920, had a number of interesting articles:

A Great Alumni Meeting

"With the object of promoting the Greater Citadel project, the Association of Graduates held a great reunion meeting in Columbia on January 21st. In the evening, in the ballroom of the Jefferson Hotel, a splendid banquet was served at which the members of the General Assembly, then in session, were honored guests. Mr. D. G. Dwight, president of the Association, presided, and addresses were made by Governor Cooper, Lieutenant Governor Lyles, Hon. T. P. Cothran, Speaker of the House, Lieutenant Colonel Barnwell R. Legge, Senator Padgett, Mr. W. S. Lee, of Charlotte, Mr. F. B. Grier, Colonel Bond, and others.

"A beautiful plaster model of the grounds and proposed buildings to be erected for the new Citadel had been sent down from Boston by the engineers, Lockwood, Greene & Co., and was on exhibition in the hall, and also a number of the architect's drawings of some of the principal buildings.

"At the meeting of the alumni in the afternoon, it seemed the concensus of opinion that the graduates ought to provide for the athletic activities of The Citadel at the new site, and Mr. W. S. Lee, chairman of the building fund committee, met the enthusiastic approval of those present when he said that he was going after a hundred-and-fifty-thousand dollar fund for that purpose."

Greater Citadel Plans Approved by Legislature

"The first appropriation made by the Legislature at the session of 1919 was for $300,000, to be paid in three installments. By the same act, the Board of Visitors was authorized to sell the present buildings and site at Marion Square and devote the proceeds to further construction at Hampton Park.

"It became apparent at once, however, that an advantageous sale of The Citadel property could hardly be made at once for several reasons. In the first place, it could not be sold with the condition of giving possession to the purchaser for some time to come, as it will be necessary to occupy the present plant until the new one is ready for occupancy. It is also very unlikely that the full value of the property could be obtained under a

Alumni Hall

forced sale. The Legislature has wisely decided at the present session to allow the Board three years in which to dispose of the property, and in the meantime, in order that the new plant on the Ashley River may be constructed without delay and made ready for the transfer of the military college, they have authorized the Board to borrow $600,000 on the credit of the State, and to sell the property at Marion Square at leisure and to the best advantage, turning the proceeds into the State treasury.

"With the funds thus made available, the Board will be able to carry forward simultaneously the construction of the main barracks building, the college building, and other auxiliary buildings needed for the complete project.

"It is expected that everything will be ready at the new site on the Ashley by the summer of 1921."

✯ ✯ ✯ ✯ ✯

In April, 1920, Mr. Lee announced that he had received subscriptions to the Alumni Fund from ninety-nine graduates amounting to $49,378, and the *Citadel News* had this item:

"The work on the buildings of the Greater Citadel is progressing satisfactorily out at Hampton Park. The large Barracks Building is now up to the third story, and begins to show how imposing this structure is to be.

"The Mess Hall and Kitchen are under one roof, but are unfinished, being occupied as a storeroom for cement and supplies.

"The extensive concrete foundations of the College Building at the south end of the parade ground are finished, and the walls will soon be going up. It is in one corner of the West Wing of this building that the corner stone will be laid on Thanksgiving Day.

"The athletic fields will be on the north side. Here, it is hoped to see before long a start made on the Alumni Building, which ought to be a structure of an importance to match the College Building facing it."

The Laying of the Corner Stone

Lay her foundations deep
In the soil where her soldier-sons sleep!
Lay them square,
Make them straight,
Take thought for the weight
They shall bear.
Lay them wide as democracy's creed,
Make them strong as democracy's need.

Build up her ample walls,
These shall be Learning's halls!

> Here shall she rear her sons
> Noble and brave and strong,
> Hating the wrong.
> Let all her rooms be bright, make them clean,
> Fashion them beautiful, befitting a queen.
>
> Raise up her tower high,
> Upstanding and straight, like the right,
> Crowned with light,
> Heaven-pointing and fair to the eye.
> By day it shall beckon afar!
> At night lift the soul to the star!

Thanksgiving Day, November 25, 1920, was a notable day in the Citadel calendar. At 10 o'clock, the Grand Master of Masons of South Carolina, Honorable Samuel T. Lanham, laid the corner stone of the Greater Citadel at Hampton Park, with over two thousand Masons in full regalia, assisting in the imposing ceremony. Mr. Orlando Sheppard, chairman of the Board of Visitors, himself a past Grand Master, presided, and the address was delivered by Governor Robert A. Cooper, another past Grand Master.

The weather was ideal, and a concourse of people estimated at 5,000, assembled to witness the exercises. The procession formed at the entrance to Hampton Park, and led by the cadet band and battalion of nearly 300 cadets, with several hundred alumni, and the imposing procession of 2,200 Masons, marched through the Park and across the parade ground to the West Wing of the college building where the corner stone was laid.

On the stand were gathered a group of distinguished citizens, the Board of Visitors, The Citadel faculty, and behind these, on raised seats, the chorus of a hundred boys and girls from the Charleston Orphan House, who sang the Masonic odes, led by the Metz Orchestra.

The exercises were interesting, picturesque, and impressive, and a perceptible feeling of solemnity and dignity pervaded the proceedings, enhanced no doubt by the expansiveness and beauty of the scene, and the imposing proportions of the structures now beginning to rise upward out of the ground.

Two other features besides the laying of the corner stone made the Citadel Jubilee on Thanksgiving Day memorable.

One was the annual football game between those ancient rivals—Carolina and Citadel. The Gamecocks came down with the prestige of a $6,000 coach, and of not having had their goal line crossed this year by a State team, and were favorites to win by at least two touchdowns. Never before had such a crowd gathered at a football game in Charleston as was seen at noon at Hampton Park, and the stands on both sides of the field overflowed with gay, colorful, and enthusiastic rooters, keyed up, when the game began, to the highest pitch of expectation and excitement.

The whistle blew for the kick-off, and then happened one of those marvelous and thrilling events which spectators are fortunate to witness few times in a lifetime. Carolina's fullback, Gressette, made a perfect kick. The oval soared high in the air, and descended far back into the waiting arms of the Bulldogs' quarterback, Jack Frost, who started for the enemy's line—eighty-two yards away. The oncoming Gamecocks closed rapidly in upon him. How far could he go before they would bring him down? Darting, winding, and twisting, to the amazement of 4,000 spectators, Frost eluded foe after foe, and finally squirmed his way completely through the enemy's line. Then began the hair-raising race for the goal line, with the ravening wolves—they were not birds now—hot on his track. Everybody held his breath. But it did not have to be for long, for Frost must have made a new record for those forty yards, and had scored a touchdown in about a minute's play!

This was the episode. What followed was a battle royal in which every member of both teams gave all he had. It was not until the third quarter that Carolina after strenuous work shoved across her lone touchdown, which with Gressette's one point for kicking goal, gave her a scant seven to six victory.

It was a great game!

The alumni banquet in the evening in the Citadel mess hall was a splendid celebration. It was probably the largest meeting of the alumni since the Semicentennial in 1893. The address of Major Mendel L. Smith met the high expectations of his hearers, and the impromptu speeches of other alumni were full of enthusiasm and suggestion.

For the information of the alumni, *Citadel News,* in its issue No. 11, March, 1921, gave a map of the city of Charleston showing the location of The Greater Citadel, and giving an account of the work to date:

"By reference to the map of the City of Charleston which is reproduced herewith, the location of the land given by the City to the State for the Greater Citadel will be seen. It consists of a tract of seventy-six acres of high land lying just west of Hampton Park. Its average elevation is about seventeen feet, the highest part, known as 'Indian Hill,' rising to about 29 feet. This is undoubtedly the highest tract of land in the city.

"Between Indian Hill and the Ashley River is a stretch of salt marsh land of nearly a hundred acres which also forms a part of the Citadel property, and may in time be filled in by dredging from the river, as was done in the case of the Boulevard development.

"Along the marsh boundary on the south and west are numbers of majestic old live oaks whose spreading branches and perennial foliage add greatly to the beauty of the landscape. Just across the river, a mile away, is historic 'Old Town,' where the first settlers landed and made their home in 1670. It was ten years later that they moved across the River to the peninsula where the present city was built.

"Two bridges span the Ashley in the neighborhood of The Citadel. In fact the Seaboard Railroad runs through the northern part of the property. A little less than a mile to the south is the highway bridge which carries the St. Andrew's Parish traffic to and from the city at Spring Street.

"A favorite drive for parties visiting Hampton Park is the mile of road which makes a circuit of the Citadel grounds passing over Indian Hill and along the border of the marsh. One part of this road, the 'Sycamore Avenue,' is particularly beautiful in summer.

"A site for the Cadet Hospital has been selected on an elevation adjacent to Indian Hill and very little inferior in height to it, and it is hoped that this building will be one of the first of the buildings to be added to the plant in the future.

"At the north end of the parade ground is the site of the Alumni Building, for which funds are now being solicited among the graduates. The athletic fields will be laid out beyond the railroad, where there is ample room for the football gridiron, baseball diamond, and running track. The tennis courts will probably be laid out just east of the parade ground next to Hampton Park.

"The following figures furnished by the engineer in charge will give some idea of the magnitude of the work at the new site: One of the first things to be done was to build half a mile of railroad track into the grounds, a spur of the Seaboard Air Line Railway, so as to have material delivered on the spot where it was needed. Four hundred and seventy-five carloads of material have come in over this track, including 10,000 barrels of cement, nearly 700,000 brick, and half a million interlocking tile.

"Over 2,200 cubic yards of concrete have been laid into the foundations and a million feet of lumber used in the wooden part of the construction. If the stucco and cement plastering were laid out on one surface, it would cover about fourteen acres.

"A person standing in the great courtyard, or 'quadrangle,' of the barracks building cannot fail to be immensely impressed with the giant size and beauty of this structure. The twenty-six great reinforced concrete

"The Courtyard of a Thousand Arches"

arches of the second-floor gallery surmounted by more than a hundred smaller arches on the third and fourth floors, give a very impressive appearance. The semi-circular arch greets the eye everywhere—in the majestic proportions of the arches on the second floor, in all gradating sizes down to the small arches in the balustrades of the galleries. The Quadrangle can literally be called 'The Courtyard of a Thousand Arches.'

"The size of the barracks building can be appreciated when it is stated that the main building of the old Citadel at Marion Square, which is 202 feet long by 131 wide, could be put up inside the Quadrangle of the new Citadel, which occupies an area of 6,100 square yards. The floor space of the rooms and galleries amounts to more than 100,000 square feet; and there are ten miles of electrical wiring in this one building."

In October, 1921, *Citadel News* reported:

"The work at Hampton Park on the Greater Citadel is now nearing completion. The Barracks Building is a magnificent structure, and is practically ready for occupation. The two wings of the College Building, with the class rooms, laboratories, and administrative offices, are nearly done, and the equipping of these buildings will soon be started. The Mess Hall and Kitchen need only some interior finishing. The Power-House and Laundry building is now in course of erection, and when done will complete the program for which funds have been provided.

"The next most urgent need is a Cadet Hospital and houses for the administrative officers, which it is hoped the Legislature will provide for at the next session, so that these buildings can be constructed next spring and the Corps of Cadets moved to the new plant next summer.

"The Citadel is looking to the alumni to start next year on an Alumni Building which will be the beginning of the athletic development so ardently looked forward to."

When a generous and appreciative city gave the State, without cost, the beautiful and spacious campus for its military college, the State was not niggardly in its provision for the buildings to be erected thereon. In two years, nine hundred thousand dollars had been expended for that purpose, and a splendid modern college plant was the result. Not that it was complete. It would have been a poor designer who had no vision of future developments. What was ready in the spring of 1922 was one great barracks building for the accommodation of 450 cadets, the wings of the college building for their instruction (the main building being left for the future), and the auxiliary buildings—mess hall, kitchens, power house, shops, and laundry—needed for the physical care of the student body.

But there were still some urgent needs not provided for.

Possibly there were some who thought that The Citadel, which had always been overly modest in its requests, might wait awhile before asking for anything more.

The administrative officers—the president, the commandant, and the quartermaster, as well as the professors, could continue to live at the old Citadel—two miles away—going up in the morning to their work at the new Citadel, and returning at the end of the day. Similarly, the cadet hospital at the old Citadel could continue to be utilized for the care of the sick—although two miles away. This was not a desirable arrangement, however, and when it was brought to the attention of the legislature in 1922, Senator Padgett and other friends in the Senate and House succeeded in obtaining authority for the Board to borrow an additional $75,000 on the old Citadel with which to put up a cadet hospital and houses for the administrative officers on the new campus.

Unfortunately, the banks did not look favorably on the loan. But now happened one of those things we cannot explain because it is not permissible any longer to believe in miracles or a "destiny that shapes our ends." Mr. J. P. Thomas, the Charleston member of the Board of Visitors and chairman of the building committee, announced that a citizen of Charleston, who requested that his name should remain unknown, had made the splendid gift of $60,000 for the erection and equipment of a cadet hospital which should be in every way modern and complete, and architecturally a pleasing addition to the group of buildings.

And thus, at the end of the decade, 1912-1922, after eighty years' association with the historic building on the Citadel Green, the institution prepared to move to its statelier mansion on the Ashley. The last Commencement exercises of the old Citadel were held on June 13, 1922, at Hibernian Hall. Senator James G. Padgett, Class of 1892, long a member of the Board of Visitors, and the valiant champion of The Citadel's interests in the legislature, made the annual address, and referred to the tinge of sadness in our rejoicing over the prospects of The Greater Citadel in the retrospect of the glorious past and the cherished memories associated with the old Citadel.

Mr. Orlando Sheppard, chairman of the Board, presented diplomas to the fifty-four members of the graduating class—appropriately the largest class up to that time in the history of the institution.

Mary Bennett Murray Hospital

CHAPTER XI
FIFTH DECADE: 1922-1932

IF a phrase be needed to characterize the special purpose which was kept constantly in view during the next decade of The Citadel's history, it might be expressed in the headline of an article in *Citadel News* for July, 1921:

RAISING THE STANDARD

"The prestige of The Citadel as a military institution is widely recognized. Its standing as a college, however, is not so generally accepted, and depends, indeed, upon conformity to certain standards which are fixed by the associations of the colleges and universities of the country.

"It is well known to a number of our graduates who have desired to pursue advanced courses at the universities and technical schools that their diploma from The Citadel has not been sufficient to give them the credit they need. This has been due to the fact that The Citadel course has not been recognized as fully up to standard; and the principal reason is that the requirements for admission are not high enough. No college at the present time can hope to be classed as standard which does not require at least fifteen Carnegie units of high school work, or the completion of a four-year high school course.

"(To the everlasting honor of its perspicacity in looking beneath the surface to find the substance, we wish to record the action of Johns Hopkins University in accepting Citadel graduates for postgraduate work ever since its reopening in 1882. Two distinguished English scholars—Dean T. P. Harrison, of the North Carolina College of Agriculture and Engineering, and President J. P. Kinard, of Winthrop College—were admitted to the graduate courses at Hopkins upon their Citadel diploma and received their degrees as doctors of philosophy.)

"In order that The Citadel may receive full credit, the Board of Visitors passed a resolution at its recent meeting in June, 1921, that steps be taken as early as practicable to meet the requirements for membership in the Southern Association of Colleges. Membership in the Association will give the diploma and degree of The Citadel the recognition which will enable our graduates to be admitted to postgraduate work for higher scholastic and professional degrees in any of the universities of the country.

"There has been another matter which has operated to retard the recognition of The Citadel as being of collegiate rank. It may seem a small matter by what names an educational institution, its officials, and its classes may be called; but, in the popular mind, an institution is practically classified by the name it uses. If it is called an 'academy,' there will be an unending struggle to show that it is not a 'prep' school. The question will arise: 'Why, if it is really a college in its work, is it called an "academy"?'

"The title of the presiding officer of the colleges of the country is almost universally the 'president,' and not the 'superintendent'—a word used for officials of different kinds.

"Worst of all, the numerical names of the classes cause a hopeless confusion in the public mind—an outsider never knowing whether a 'first-classman' is in his first year at college or his last. Of course, these titles at The Citadel were originally adopted following the peculiar system in vogue at West Point and Annapolis, rather than the universal practice of the colleges.

"It is not likely that The Citadel will ever abandon or greatly modify its military system, which has proved so effective in training disciplined citizens who also became competent soldiers when they were needed; but The Citadel must rank as an educational institution, doing college work for the State, and not as a purely technical school whose business is solely the training of officers, as at West Point. The terms which it uses ought, therefore, to be those in general use in the colleges. Some years ago, the name 'academy' was abandoned, and the word 'college' substituted. The Board has now changed the title 'superintendent' to 'president'; and the classes will be known hereafter by the terms universally understood in the college world: senior, junior, sophomore, and freshman."

One of the first objectives in raising the standard was, as stated, raising the entrance requirements. At the beginning of the World War—that is, for the session of 1916-1917—these requirements were raised by the colleges of the State from ten to twelve units. There was a consequent falling off in the freshman classes of all the colleges which enforced the standard. At The Citadel, the superintendent reported: "Owing to the backward condition of the high schools, many of them can not get their graduates admitted to college on a certificate of three years' work, and a large number of applicants for admission to the freshman class at The Citadel were rejected on the entrance examinations."

"An examination" had been the time-honored method by which applicants had been admitted to The Citadel. But admission by this route was the hard and stony up-hill way—a "certificate," showing so many "high school units" was the paved highway that the high schools and colleges of lower grade were seeking to construct. Finally, about the year 1922, when —owing to the phenomenal development of the State high school system—it became possible in South Carolina for the colleges to require four years of preparation in an accredited high school, covering at least fifteen units, for admission, the millennium in education was supposed to have arrived.

Under this system, it must be frankly confessed, it was quite possible for a student to enter college without being able to name the president of the United States or to add common fractions—provided that he had been "exposed" to education for four years, it was not necessary that it "had taken," and, indeed, the colleges found out that many were "immune."

Naturally, any effective educational system cannot grow up in one or two years; and it must be said to the credit of the high schools that they realized very often that they were engaged in the impossible task of trying sometimes to make silk purses out of sows' ears. They would honestly label their output as "college material" or the reverse, leaving the colleges the responsibility of accepting those in the latter class. A college that needed students would, of course, be willing to assume responsibilities of this kind.

An amusing incident occurs to the writer in this connection.

The examination for vacant scholarships in the State colleges is held in July each year. This is not an examination for admission—since all the candidates are supposed to be graduates of high schools—but is merely a competitive test on elementary subjects in order to determine the best prepared candidate.

One summer, in a certain county, there were three candidates, all of whose grades on the examination were so nearly utter failures that the college would not recommend any one of them for appointment. In the fall some time after the college had opened, a freshman came to the president's office to inquire what had happened about the vacant scholarship in this county, and, upon being questioned, acknowledged that he was one of the unsuccessful contestants.

"Well," said the president, "with such low grades as you made on the examination, how is it that you are at The Citadel?"

"Oh," was the smiling reply, "I came as a pay cadet, and entered on my high school certificate." And maybe he made good.

And just here it might be well to recall what a wise South Carolina pedagog has to say on this subject. In his quaint and delightful address to the graduates of The Citadel at the homecoming banquet on October 20, 1926, Elliot Crayton McCants gave this advice:

"All teachers should quit that bad habit known as 'passing the buck.' At present, the colleges attribute their failures to the poor work of the high-school teachers; the high school lectures the grammar school; the grammar school berates the primary department; and the primary department writes a note to the mother. She, poor woman, sighs and remarks resignedly, 'Well, your father's folks never did have much sense, anyway.'

"It is time that every teacher, all along the line, accepted his personal responsibility. If a student knows nothing, it is your job and mine to teach him something."

Beginning with the first session at the Greater Citadel, the standard required by the Southern Association of Colleges was put into effect, and was maintained for two years before application was made for admission to the Association.

The Citadel's credentials were investigated by Professor Hand, of the University of South Carolina, and her application was sponsored by Professor C. A. Graeser, of the College of Charleston, a member of the Commission on Higher Education.

On Friday morning, December 5, 1924, at the morning session of the Association, held in the Hotel Chisca, Memphis, Tennessee:

"The Executive Committee recommended for membership in the Association the following institutions of higher education: The Citadel, Charleston, South Carolina; Furman University, Greenville, South Carolina; and Agricultural and Mechanical College of Texas, College Station, Texas. On motion, the vote of the Association was cast for these institutions, and they were declared elected to membership."

And the above date must be recorded as marking one of the important milestones in the progress of The Citadel.

Slowly The Citadel had progressed in its struggle for recognition as an institution of higher education.

It was proud to be received into the select sisterhood of Southern colleges.

But it must be said that at no time in the past had it suffered from any inferiority complex—to use a modern term. Year after year it had published in the back of its annual catalog a list of its alumni, with their present occupations and addresses, saying to the world, "These are my children!" and being willing to submit to the ancient standard, "By their fruits ye shall know them"—leaving to the public to decide whether her confidence had been justified in the rank her sons have taken in the world.

In another matter The Citadel's policy had been slow to alter.

From its beginning in 1842 down to the year 1903, its four-year curriculum had been inflexible. According to all modern theories of education, this was altogether unsound. This is to ignore special aptitudes and tastes. As one expert put it, "If a boy is not interested in algebra, but loves horses, let him study about horses." Perhaps so. The boy, however, does not wish to get his semester credits in a livery stable, but in college. It is true that some versatile universities may offer equine and asinine courses, but the great majority of colleges must limit themselves for economic reasons to a more limited field. That The Citadel offered only one rigid course, requiring every student to take four years of mathematics, four years of science, four years of English, four years of modern languages, four years of history and social sciences, and four years of "military science"—on which was superposed four years of practical instruction in how to regulate one's daily life—seems indefensible and, at the present time, incomprehensible. One result, beyond doubt, was that at the end of the session only the fittest had survived.

It would be profitless to argue that the hard road to learning is better than the line of least resistance. Colleges, like individuals, will continue to vary—some giving as much and some as little time as possible to the "broad foundation" before letting the student specialize. The Citadel's policy of one required course, with no electives, was at one extreme; the policy of some others, allowing specialization from the very beginning, with only the high-school course for foundation and background lay at the other extreme. There are indications that these extremes are now beginning to converge—making the first two years of the college course, the junior college—cover the general foundation, after which the student elects and pursues a chosen specialty.

The adoption of electives in the senior class was made at The Citadel, as we have seen, in 1903. Even then, it was a limited, three-way elective, the seniors being required to choose a major study in either civil engineering, the sciences, or a literary course. Certain other required courses were still taken in common. It was not until 1916 that the elective system was extended to the junior class, giving two years' work in some special chosen line. In 1924, an important extension of the electives was made by the introduction of a fourth direction which students might pursue in the last two years of study. In the spring of 1909, President Eliot, of Harvard, was a visitor in Charleston, and by invitation spoke to the cadets about the experiment at Harvard of college courses in business management.

He pointed out that commerce, finance, manufacturing, and "big business" in general offered fully as large a field for talent, and was just as potent in "educational value," as engineering or the professions, and was therefore equally deserving of college study and recognition. During the next ten years the Cambridge example was followed by many other institutions of higher education.

In 1924, Dr. G. L. Swiggett, a specialist in this field in the Bureau of Education at Washington, was invited to deliver the Commencement address at The Citadel, and to advise about the introduction here of an elective course in business administration. It was the course outlined by Dr. Swiggett which was adopted, and has been continued with certain important modifications to the present time, offering a popular course for a large percentage of the students.

A few years later, in order to give the special training required of teachers in the public schools, elective courses in education and psychology were established, so that, at the present time, 1932, cadets at The Citadel who have completed the sophomore class can elect specialized courses in civil engineering, physics, and electrical engineering, chemistry, pre-medical chemistry-biology, or business administration, all leading to the degree of bachelor of science; or they may pursue advanced studies in English, history, and the social sciences, or modern languages, leading to the degree of bachelor of arts—the latter degree being first awarded in 1925. Higher mathematics, psychology, education, and geology are courses offered as minors in all elective courses.

This development of the elective system could never have been attained with a very small student body. With an enrollment of less than 150, and a senior class of no more than twenty-five members—a condition that existed for many years—the per capita cost of specialized tuition would have been prohibitive. A chair of Sanskrit might be maintained at Harvard from its endowment if only three or four students desired to study it, but not at a State institution dependent upon the legislature for an appropriation.

Of the 110 graduates in 1932, twenty-three took the bachelor of arts courses; twenty-six, civil engineering; twenty-two, science; and thirty-nine, business administration—these last three groups receiving the bachelor of science degree.

✦ ✦ ✦ ✦ ✦

During the first half of the decade, 1922-1932, the average total enrollment was just under 400; and during the last half, 675—reaching its peak in the years 1927 and 1928, when it reached 722.

This third decade of the twentieth century will long be remembered for the astonishing popular interest in education.

As in other matters, the American people lived up to its reputation as a democracy run on impulses by embarking upon an era of spending (for "educational progress"), that well-nigh assumed the proportions of an orgy. Cities and towns spent millions upon buildings and equipment for their grade and high schools, and even country districts erected in the woods examples of classical architecture that would have been an ornament to any city.

Consequently, in a few years, there was a swollen stream of applicants surging from the high schools into the open doors of the colleges.

When it is remembered that the dormitory capacity of the Greater Citadel when it was first occupied in 1922 was only 430 cadets, it is evident that when the enrollment reached over 700, something must have happened in the meantime. What it was can best be told in the words of the report of the chairman of the Board of Visitors to the legislature for the year 1926-27:

"November 30, 1927.
"To the State Superintendent of Education, Columbia, S. C.

"Dear Sir: I have the honor to submit herewith for the information of the General Assembly the following report of the affairs of The Citadel for the fiscal year ending June 20, 1927.

"The enrollment for the year was 580, of which number 269, or 46%, were new students. As the capacity of the main barracks is less than 450, it became necessary to do some temporary crowding until the new barracks could be completed.

"Of the total enrollment, the great proportion—86%—were from South Carolina, ninety-three cadets being from nineteen other states.

"The commencement exercises were held on June 3d to 7th, being opened on the first date with a review of the corps of cadets by Major General Charles P. Summerall, Chief of Staff of the United States Army. The graduating exercises were held on the 8th, the annual address being made by Gen. Lawrence D. Tyson, U. S. Senator from Tennessee. Diplomas were awarded to sixty members of the senior class—the largest graduating class in the history of the institution.

"The most noteworthy event of the past year was the building and occupation of the Andrew B. Murray Barracks, a splendid dormitory building rivaling in size the main barracks, and making it possible almost to double the student capacity of the Military College. This building was opened with appropriate public exercises on April 22d, at which an address was made by Mr. Walter B. Wilbur, of Charleston, who, in the course of his remarks, paid a fine tribute to the character and public spirit of Mr. Murray.

"The outstanding importance of this event was immediately signalized by the applications of a record-breaking freshman class of 367 men. But for the new barracks the number of freshmen would have been limited to less than a hundred.

"The striking increase in the enrollment at The Citadel is shown in the table below:

"Session of 1924-1925, 313 cadets.
"Session of 1925-1926, 438 cadets.
"Session of 1926-1927, 580 cadets.
"Session of 1927-1928, 722 cadets.

"It will be seen that Murray Barracks has enabled the College to more than double its enrollment in three years.

"When The Citadel was transferred from Marion Square to the new site at Hampton Park five years ago, accommodations as to dormitories, classrooms, mess-hall, and hospital facilities were all based on a student enrollment of 450. In the larger plans for a Greater Citadel, provisions were made for the future construction of other buildings as need would arise. On the basis of 450 cadets to begin with, only the main central barracks building was constructed, leaving space for two additional buildings, one on each side, to be built when The Citadel grew. At the end of three years this need became apparent, and it was then that the public-spirited citizen of Charleston came forward with his proposal to give $150,000 for this purpose if the legislature would appropriate a like amount—a proposition the State did not hesitate to accept. As a result, the 'Murray Barracks' stands today literally in the concrete, and with the Main Barracks furnishes accommodations for 750 cadets.

"At the same time it has been necessary on two separate occasions to enlarge the Mess Hall.

New College Building

"In the Greater Citadel plans, the architects provided for a Main College building with two wings. Five years ago, with less than 400 cadets, the wings sufficed for the needs of the institution. Today, the facilities for the instruction of the student body are not adequate, and the construction of the Main Building is imperatively needed.

"The plans for this imposing central structure call for a Memorial Hall, the administrative offices, a beautiful and spacious Library, and a number of additional class-rooms and laboratories. Estimates of the cost of this building have been made by the architects and indicate that about $250,000 will be needed for the purpose.

"It is my duty as chairman of the Board of Visitors to present to the Legislature in the clearest and most earnest terms the absolute need of the Military College for these facilities for instruction and administration, and to ask for the requisite appropriation at the next session of the Legislature in order that the institution may not be set back in its work and development.

"The Act of the Legislature permitting the State colleges to use tuition fees for building programs has been of great practical value during the past two years.

"When it was found that the new Murray Barracks could be constructed at a cost under the estimated amount, and that there would be a substantial balance from the joint Murray and State fund, the Board immediately took

Barracks—Greater Citadel

up the consideration of building the much-needed quarters for the principal administrative officers of the college.

"It is desirable for all the professors to live on the campus; but it is particularly important that the President, the Commandant, the Quartermaster, and a few others live on the grounds and be in contact with the affairs of the college at all times.

"The fortunate economy realized in the building of Murray Barracks opened up the possibility of providing homes for those officers which the Board was glad to seize.

"The architects were engaged to draw plans and specifications for a President's House, and an Apartment House for four other officers. The consent of Mr. Murray was readily obtained for the expenditure of the balance of his contribution in this way. When the bids were opened, however, it was found that the lowest bid was in excess of the amount available, and the only way in which the contract could be awarded was for a transfer of student-tuition fees to be made for this purpose, consent to which was promptly made by the Contingent Fund Committee, thus enabling the Board to go ahead with this most important building program.

"These two useful and beautiful additions to the college plant were completed during the summer and were occupied by the five senior faculty officers at the beginning of the present session.

"Another instance of the value of the provision for the use of tuition fees for building purposes arose in the summer when it became evident that the enrollment of new students would be unexpectedly large, and that the Mess Hall would have to be enlarged by an extension of fifty feet in order to accommodate the corps of cadets. With the tuition fund available, it was possible, at short notice, to build this necessary addition and have it ready for the cadets on the opening day—Sept. 20th—a most important matter.

"A provision for meeting an emergency of this kind, when it occurs, is certainly much better than the chance method of trying to anticipate possible wants a year in advance. It relieves the budgets of the colleges of minor but necessary items, and sometimes permits the governing boards to plan even considerable improvements, adding thereby to the usefulness and value of the State's property.

The Old Citadel

"On account of the limited accommodations for officers at the new Citadel, the Old Citadel at Marion Square is of necessity still used as living quarters for many of the professors—eighteen married officers and their families, and six bachelor officers, having apartments in the buildings at the present time.

"The expense of repairs and up-keep of these apartments has been largely borne by the officers living in them. The Boy Scouts are allowed to use several of the rooms on the ground floor of the King Street Extension for their headquarters and assembly room. The Medical Detachment of the National Guard also has the use of a number of rooms in the Meeting Street wing. For several years past, permission has been given to the Charleston County Agricultural Society to use the old barracks building for a County Fair for a week in the fall. Thus, while the Old Citadel has been abandoned as a college plant, it has been by no means deserted. For some years to come it will continue to serve the useful and essential purpose of housing the faculty, until the new plant on the Ashley River shall be developed to the point where the faculty can be housed on the campus.

"The Old Citadel has a frontage on Marion Square of 630 feet, and a depth of 147 feet on King and Meeting Streets. Its location is in the very heart of the City. The intersection of King and Calhoun Streets where the Francis Marion Hotel stands is probably the busiest street corner in Charleston, and the center of the retail business and of the amusement houses of the city. The Citadel property is potentially one of the most valuable pieces of real estate in Charleston, and there should be no thought of sacrificing it at the present time when real estate values are generally low.

(Signed) "JOHN P. THOMAS, Chairman."

Although the Board of Visitors took as its first definite objective the construction of the main college building, and put an item in The Citadel budget for this purpose, the legislature which met in January, 1928, was not favorably disposed, and the attack was shifted to a bill to authorize the Board to borrow $250,000 on an amortization plan, pledging future tuition fees for payment. This, however, failed.

In the fall of 1928, the chairman, in his report to the legislature, reverted to this plan. He said:

"It is my duty as Chairman of the Board of Visitors of The Citadel to call attention of the General Assembly again to the urgent and pressing need for the Main College Building. If the institution is really to become the great college which its mission calls for, this building should be constructed at once."

He proceeds: "In financing their building programs, the governing boards of the colleges throughout the country have sought to attain their ends by various methods. These methods may be classed under three general heads:

"1. A State bond issue.

"2. Direct legislative appropriations, or special millage in the tax levy.

"3. Private financing, by campaigns or benefactions.

"The first of these methods has been adopted in a number of States, but was decisively defeated in South Carolina a few years ago when a Constitutional Amendment for that purpose was submitted to the people.

"The experience of the past two years gives little hope that sufficient funds can be obtained from current appropriations.

"As to the third, The Citadel has been particularly fortunate in receiving several notable benefactions, which have provided for our beautiful and well-equipped Hospital and the splendid new Murray Barracks. Funds from the same source have contributed largely in the additions of the President's House, and the Officers' Apartments on the campus. In these latter buildings, the cost has been partly borne by funds from tuition fees. The construction of Alumni Hall is due to the voluntary subscriptions of the graduates. But benefactions cannot be counted upon even when most needed. Therefore, the proposal of the Board to utilize tuition fees to amortize a loan seems to be the only feasible plan yet devised for continuing our needed building program."

Lack of success in achieving their purpose did not deter the Board from renewing their application to the legislature again next year. This time, the loan plan was given up, and a direct appropriation asked for. The report of the chairman, October 15, 1929, says:

"During the past summer, part of the Old Citadel has been fitted up so that three more professors can be provided with quarters. There are now 32 families living in these buildings. At the new Citadel, accommodations for only five officers have been constructed up to the present time.

"While the desirability of building apartments for the faculty on the campus of the new Citadel has been recognized and kept in mind, the Board has not been urgent in its insistence on an appropriation for this purpose, preferring to use the old buildings at Marion Square until more imperative needs are provided for.

"The great and most pressing need at The Citadel at this time is the Main College Building.

"The general public does not seem to be aware that if the enrollment at The Citadel should be limited to its present number, 720, the need of a building program would still be an urgent matter. The impression exists that we now have practically a complete plant for the number of students in attendance. Such is not the case. It may seem strange to the uninformed that the Main College Building at The Citadel has not yet been constructed, but such is the fact. It was planned by the architects in 1920, and is designed to be the most imposing structure on the grounds. Its erection was deferred because at that time the size of the corps was less than half what it is today, and the two wings of the College building were then sufficient for our needs. They are now altogether inadequate. An enumeration of the deficiencies in our present equipment will make the point clear: A well-equipped library is counted among the most important of modern educational agencies, and is considered indispensable in a college that lays any claim to scholarship and learning. A library, however, is one of those things that requires housing and room, and must wait upon a proper building. A complete and commodious home for a library should be one of the first considerations in a building plan, and it might with propriety form the central and conspicuous feature of the new Main Building.

"Besides a number of administrative offices, at least sixteen classrooms should be provided in this building so that the ground floor of Murray Barracks which is now entirely given over to academic classes, may be used for its legitimate purpose of dormitory accommodations.

"Estimates of the cost of this Main College Building have been made, and the Board feels impelled to urge with all earnestness a special appropriation for this purpose."

Next year, the Board's objective was finally won—how, is explained in the following extracts from the chairman's report to the legislature, November 21, 1930:

"During the present year, all tuition fees when collected have been deposited with the State Treasurer as required by a provision of the Appropriation Act which reads as follows:

" 'Provided, that for the year 1930, all tuition fees and other undesignated fees collected by The Citadel shall be paid into the State Treasury monthly, between the 1st and the 10th of each month. This fund shall be held in trust by the State Treasurer to the credit of the institution, and shall be paid out and expended for permanent improvements or other purposes on the order of the Board of Trustees subject to the approval of the State Finance Committee.'

"About a year ago, the Board of Visitors passed a resolution raising the tuition fee of cadets from outside the State from $40 to $100 a year. Consequently, during the present session, which began on Sept. 13, the cadets from other States (about 170 in number) are paying tuition at this higher rate, and a much larger income is made available from this source for new buildings and permanent improvements.

"Last summer, the Board decided, in view of the unusually favorable price conditions for building construction, to erect the Main College Building at the New Citadel. This building has been much needed for several years, but although an appropriation has been requested each year in the budget, the Legislature has declined to approve a direct appropriation, but has adopted a policy of permitting the State colleges to make use of tuition fees for building purposes.

Bond Hall — Main College Building

"With the approval of the State Finance Committee and the Sinking Fund Commission, a contract was entered into for the erection of the Main College Building at a cost of $123,650, to be paid by fees already collected and future fees pledged for the next few years. This price is considered to be remarkably low, and the construction of the building—now in progress—will in all probability be completed by the end of the present session next June, and will add to the plant of the New Citadel one of its most important and impressive units."

This crowning achievement marks, indeed, one of the noteworthy steps in the progress of the Greater Citadel, and the Board could have been well content to rest for some years to come upon its accomplishment.

But in the spring of 1931, Senator Hammond reported to the Board that an additional loan to The Citadel of $22,000 had been approved by the Sinking Fund Commission, and was available for the construction of a duplex house—providing a home for the dean and the commandant on the campus.

This building, which conforms to the prevailing type of Spanish-Moorish architecture, is situated under the majestic live oaks between the president's house and the officers' apartments, and adds greatly to the general effect.

And thus, at the close of the first decade of the new Citadel, we find it provided with a material equipment adequate to carry it forward in its mission of service to the State. And its alumni, in particular, can rejoice that their efforts have contributed in no small degree to this accomplishment.

Academic Degrees

The Citadel was slow in assuming some of its privileges as a recognized college.

In the matter of degrees, we have seen that it was not until 1900 that the Board of Visitors asked the legislature for authority to grant the degree of bachelor of science. Even then, for some years, this degree was limited to graduates who attained very high honors on the four-year course—a sort of Phi Beta Kappa society.

Very justly, the degree was extended to all graduates in 1905.

In 1912, the act was amended to extend the power of the Board to grant the degree of civil engineer. This was not an honorary degree, but a postgraduate degree earned by responsible engineering work and the submission of an acceptable thesis.

Again, when the elective system had differentiated the students into two groups with fairly distinct objectives, the bachelor of arts degree was added to The Citadel's list of credentials.

Finally, in 1929, in order to recognize in a fitting way the public services of some of its alumni, the degree-conferring power of the Board was extended to honorary degrees. At the Commencement, in June, 1929, the

Board awarded the degree of doctor of laws to Mr. R. O. Sams, Class of 1861—oldest living graduate, and veteran educator—and Mr. Orlando Sheppard, Class of 1865—former chairman of the Board, and distinguished lawyer and citizen.

Since that time, the degree of LL.D. has been conferred upon the following distinguished alumni:

Honorable Joseph W. Barnwell (ex-cadet), Charleston.
Captain Ellison A. Smyth (ex-cadet), Flat Rock, N. C.
Dean T. P. Harrison (1886), Ph.D., Raleigh, N. C.
President James P. Kinard (1886), Ph.D., Winthrop College.
Reverend John Lake (1891), Canton, China.
Right Reverend A. S. Thomas (1892), D.D., Episcopal Bishop of South Carolina, Charleston.

The degree of doctor of letters has been awarded twice—to two of the alumni who have earned high reputation in the field of letters as novelists and historians:

Professor E. C. McCants (1886), Anderson, S. C.
Mr. William E. Woodward (1893), New York City.

The one doctor of science degree so far awarded was conferred at the recent Commencement on June 3, 1932, on William States Lee (1894), LL.D., hydro-electric engineer of international renown, New York City.

It will be observed that all of the above were Citadel men. In the original act conferring upon the Board the power to grant the bachelor of science degree, the words "upon graduates of the Citadel" naturally occurred. In the various amendments to this act, these words remained; so, when in 1932 it was desired to confer the doctor's degree upon Colonel Robert R. McCormick, editor of the Chicago *Tribune,* who made the Commencement address for The Citadel on June 3, it was found that the Board did not have authority to do so—Colonel McCormick not being "a graduate of The Citadel." The difficulty, however, was surmounted when Senator Hammond got the legislature to simplify the act by giving the Board general power to "grant degrees." One of the interesting features of the 1932 Commencement, therefore, was the investment of Dr. McCormick and Dr. Lee with the hoods of their office, showing the blue and white colors of The Citadel.

𝟏 𝟏 𝟏 𝟏 𝟏

HOME-COMING DAY

Alma Mater—"Mother dear!"—
From far and near
Her sons come back on this glad, festal day,
Her praise to sing, and homage pay.
In days gone by, she sent them forth,

Glad in the strength of youth,
Seekers of truth,
And eager for the strife.
In all the walks of life
She saw them take their place—
Workers in the world, and pioneers,
Crusaders for the right—
These have her praise,
                And those her tears
Who valiantly have fallen in the fight,
Content the world should profit by their sacrifice.
None shall be more precious in her eyes
Than these, in all the years.
    And ye, her living sons—go forth, be strong,
Cease not to battle wrong!
Assured of Alma Mater's trust, whate'er betide.
Your happiness shall be her constant prayer,
And your success her pride!

Coming back home—the words have a peculiar charm—even when they mean returning to academic walls. For every college worthy of the name gives its sons ideals and associations connected not only with its classrooms and laboratories, but with the whole vivid life which a student leads—in the dormitories, on drill, on guard, in camp, on the athletic field, and in the social functions which form his week-end relaxation.

And so, periodically, colleges ought to have home-coming days.

The first Home-Coming Day of the Greater Citadel was held on October 25, 1924. Hundreds of the alumni—old men, middle-aged, and young men—many from distant states—came to the celebration.

In order to register, they reported at the sally port of the new Citadel, entering through the same old iron gates that every one of them had gone in and out of many hundreds of times in their cadet days at the old Citadel.

The morning was spent on the campus, with many interesting reunions and talks of old times, and at noon everybody gathered on Indian Hill under the live oaks for an al fresco barbecue lunch.

Shortly after one o'clock the crowd began to wander in groups towards Hampton Park, where the chief event of the day was to take place. This was the Furman-Citadel football game, in comparison with which all other features of Home-Coming Day (and there were several others of noteworthy interest) paled into insignificance.

On this battlefield of the gridiron, two teams of stalwart warriors were to battle for the honor and renown of their Alma Mater, and to perform exploits that would put their names in big headlines in the morning papers. This was the opportunity, too, when the alumni could wear their college colors and show their loyalty to the old school.

The Purple Hurricane came down from Greenville with a number of stars of the first magnitude, and outweighing the Bulldogs a dozen or more pounds to the man, so that what the Cadets lacked in avoirdupois had to be made in "avoirdésprit"—which may be French for what is known on The Citadel campus as "Bulldog spirit."

The contest was an epic for the sports writer.

During the first two quarters "the Hurricane roared and blew, and the Bulldog reared and tore" with definite results. In the third period, the Cadets made a splendid drive into Furman territory which took the ball to within ten yards of their goal line. But here the defense stiffened, and the ball was lost on downs, and Furman kicked out safely.

Were the Bulldogs disheartened? Not much.

Says the sports writer: "The Blue and White, directed by the incomparable genius of Teddy Weeks, started again.

"Up to this time, Weeks had plugged at the Furman line. But now the uncanny Teddy let loose a few bewildering forward passes that set the Furman backs wild. A pass to Ferguson netted seventeen yards and a first down. Before Furman had settled, Teddy shot another to 'Firpo' McFarland on Furman's ten-yard line. The four thousand people in the stands were in an uproar, and Furman was visibly worried. Then Youngblood circled right end for eight yards, and two more line-bucks put the oval six inches from the line. And then Carl Hogrefe, that fighting son of Anderson, was elected to take the ball for the final try. Furman's left side met the plunge—but the ball was over."

And so was the game—as it turned out later.

Of course, at the banquet in the mess hall that night the alumni were happy, and applauded all the speeches, particularly the delightful literary address of J. Wilson Gibbes, Class of 1886, on "The Citadel of the Eighties," and the inspiring address of Major W. D. Workman on "The Citadel in the World War."

At ten P.M., the "old grads" went over to Alumni Hall to join the happy dancers at the Home-Coming ball.

And this is a good program for any Home-Coming day.

Not much has been said about athletics in these reminiscences of The Citadel. Mr. George Rogers, Class of 1910, one of the best informed of the alumni on the subject, has published in the Charleston press a complete and interesting history of football at the Military College; and it is hoped that he may be induced to extend his account to include the other activities.

The Blue and White teams have lost as well as won many battles on the field of sport; but it is their pride that they have always given their rivals the best fight of which they were capable. Above all, their directors and coaches have constantly held up to them Grantland Rice's admirable admonition:

> "For when the One Great Scorer comes
> To write against your name,
> He writes—not that you lost or won—
> But how you played the game."

✓ ✓ ✓ ✓ ✓

In the fall of 1930, General Charles P. Summerall ended a four-year detail as Chief of Staff of the United States Army, and in March, 1931, went upon the retired list of the Army.

Congress, in recognition of his distinguished record in France during the World War, and of his outstanding services as Chief of Staff, conferred upon him at his retirement the preëminent distinction of promotion to the rank of general, an honor which has been bestowed upon only eight military commanders in the whole history of the country, beginning with Washington.

In retiring from the military service to the ranks of citizenship, General Summerall's recognized character, ability, and administrative talents, naturally brought him many offers of high positions in civil affairs, and the Board of Visitors of The Citadel was fortunate in his acceptance of their offer of the presidency of The Citadel—a position for which his lifelong training peculiarly fits him.

General Summerall entered upon the duties of this position at the opening of the session in September, 1931—Colonel Bond, the former president, assuming the position of dean of the college.

That the military prestige of The Citadel has been immensely enhanced goes without saying. The public may also rest assured that under his administration its character as an institution for the education of young men of the State and nation in the highest elements of citizenship will be maintained.

It is worthy of note in this connection that a few years ago, when a vacancy occurred in the presidency of the Virginia Military Institute, another distinguished military officer was called to the position—Major General John A. Lejeune, who made a brilliant record in France in command of the Division of the United States Marines.

It is of peculiar interest that these two State military colleges, whose genesis, history, and ideals have so much in common, should now be presided over by two of the outstanding soldiers of the World War, under both of whom many graduates of both institutions served in France.

At the annual football game between these two premier military colleges, in the fall of 1931, these two generals met in friendly rivalry on the field of sport at Lexington. To give a real war-time flavor to the contest, General Summerall flew up from Charleston to the famous town in the Valley; but finding no landing field either there or at Staunton, his pilot made a forced landing in a convenient cow pasture, and the journey was completed in a Ford. General Summerall arrived too late for the review which General Lejeune planned for him, but—running true to form, he was not too late for the battle.

Appropriately enough, the two cadet teams fought each other in a thrilling struggle to a draw.

General Summerall's nation-wide reputation brings him many calls for addresses in all parts of the country; and these occasions all serve to make The Citadel more widely known. He had been president of the college only a few weeks, when, at a First Division reunion in New York, he received, without solicitation, five scholarships to The Citadel of a value of $550 a year each. These will be known as the First Division A.E.F. Scholarships. We record here the names of the young men appointed in the summer of 1932 to fill them: Wilfrid A. Carr, Brooklyn, N. Y.; Donald McCree, St. Paul, Minn.; MacDonald Lowe, Highland Park, Ill.; Allan L. Leonard, Los Angeles, Calif.; Arthur William Ferguson, Bellaire, Ga.

May these be the forerunners of a long line of young men who shall come from other states to get their college training along with our South Carolina boys in the historic Citadel.

¶ ¶ ¶ ¶ ¶

And now, in bringing to a close this chronicle of our Alma Mater, may the writer say just a few words to his fellow alumni at parting.

Even the casual reader of these pages must have perceived how large a part the graduates have played in the history of The Citadel. It was chiefly by their efforts, with the aid of their friends, that the institution was reopened in 1882; and in the fifty years since that time, it has been largely through the influence of the alumni—men held in public esteem —that it has weathered the storms which often threatened to destroy it.

Seemingly now it holds a high place in popular favor. But it may not always be so. If, in the future, its existence shall again be in danger, it may depend on the character of its alumni whether it shall survive.

The burden of taxation for education is sure to be a matter of serious discussion by taxpayers and legislators. It may be that the scholarships in

State colleges will be abolished. If so, The Citadel will earnestly look to private benefactions to help the ambitious boys who cannot pay their way.

At present there are at The Citadel only two endowed scholarships for South Carolina boys—both of which are gifts from two beneficiary cadets who realized the supreme value of an opportunity in their own lives. It is a pleasing thought that in all the years to come, as long as The Citadel lasts —long after the donors have passed beyond (and one of them has already gone)—there will be recurring examinations to fill vacancies in the William States Lee and the James R. Crouch scholarships, founded in the year 1924—enduring monuments to the donors and a living force continuing through the ages.

APPENDIX A
Chairmen of the Board of Visitors
1842-1865. General James Jones.

Interregnum, 1865-1877. In 1865, after the close of the two Academies, Hon. R. J. Davant, and General James Conner served for a few months as chairmen, but they had no functions to perform, since the institutions were not reopened.

1877-1898. General Johnson Hagood.
1898-1915. Colonel C. S. Gadsden.
1915-1916. Colonel W. W. Lewis.
1916-1925. Mr. Orlando Sheppard.
1925-   Mr. John P. Thomas.

APPENDIX B
Superintendents of The Citadel
(In 1921, the title of the presiding officer was changed to President.)
1. Captain W. F. Graham, 1843-1844. (Died in office.)
2. Major R. W. Colcock, 1844-1852.
3. Major F. W. Capers, 1852-1859.
4. Major P. F. Stevens, 1859-1861.
5. Major J. B. White, 1861-1865.
    Occupation by United States troops, 1865-1882.
6. Colonel J. P. Thomas, 1882-1885.
7. General Geo. D. Johnston, 1885-1890.
8. Colonel Asbury Coward, 1890-1908.
9. Colonel O. J. Bond, 1908-1931.
10. General Charles P. Summerall, 1931-

APPENDIX C
Officers of the Association of Graduates
The Association was organized Friday, November 19, 1852. It was inoperative during the War of Secession and during Reconstruction times. It was reorganized December 13, 1877.

Presidents
1. C. C. Tew, 1852-1853.
2. J. L. Branch, 1853-War of Secession.
3. Johnson Hagood, 1877-1898.
4. J. P. Thomas, 1898-1906.
5. O. J. Bond, 1906-1916.
6. T. P. Lesesne, 1916-1918.

Secretaries
1. J. P. Thomas, 1852-1853.
2. P. F. Stevens, 1853-1855.
3. J. J. Lucas, 1855-War of Secession.
4. C. I. Walker, 1877-1887.
5. O. J. Bond, 1887-1906.
6. John W. Moore, 1906-1916.

224   THE STORY OF THE CITADEL

Presidents

7. D. G. Dwight, 1918-1920.
8. J. R. Westmoreland, 1920-1924.
9. Geo. C. Rogers, 1924-1926.
10. W. S. Lykes, 1926-1927.
11. John W. Moore, 1927-1929.
12. M. A. Pearlstine, 1929-1930.
13. F. P. Sessions, 1930-

Secretaries

7. C. L. Hair, 1916-1918.
8. T. P. Lesesne, 1918-1920.
9. E. H. Poulnot, Jr., 1920-1924.
10. S. M. Sanders, 1924-1926.
11. D. S. McAlister, 1926-

APPENDIX D

List of Professors of Military Science and Tactics

1883-1886. 2d Lieut. E. M. Weaver, Jr., Second Artillery, U.S.A. Graduate of West Point. Afterwards made Brigadier-General, Chief of Coast Artillery, during the World War.

1886-1887. 2d Lieut. Albert L. Mills, 1st Cavalry, U.S.A. Graduate of West Point. Severely wounded in Spanish-American War, losing one eye. Became Superintendent of West Point, and Brigadier-General.

1887-1890. 1st Lieut. Charles H. Cabaniss, 18th Infantry, U.S.A. Graduate of West Point, 1874. Retired 1891.

1890-1893. 2d Lieut. John A. Towers, 1st Artillery, U.S.A. Graduate of West Point. Was the first P.M.S.&T. to be also Commandant. Died of tuberculosis at his home in Anderson, S. C. March 23, 1893.

1893-1897. 2d Lieut. John M. Jenkins, 5th Cavalry, U.S.A. Graduate of West Point. Class of 1887. Afterwards became Brigadier-General.

1897-1898. 1st Lieut. John B. McDonald, 10th Cavalry, U.S.A. Graduate of West Point, 1881. Afterwards became Brigadier-General.

1898-1902. Captain J. Willis Cantey, Graduate of The Citadel, 1891. No Army Officer was available during this time.

1902-1904. Captain Geo. H. McMaster, 24th Infantry, U.S.A. Graduate of West Point, 1892.

1904-1908. Captain William H. Simons, 6th Infantry, U.S.A. Graduate of The Citadel, 1890. Wounded in Spanish-American War. Was colonel at time of his death in 1917.

1908-1912. 1st Lieut. William St. J. Jervey, 10th Infantry, U.S.A. Graduate of The Citadel, 1894.

1912-1915. 1st Lieut. Jesse Gaston, 10th Infantry, U.S.A. Graduate of West Point, 1903.

1915-1917. 1st Lieut. Enoch Barton Garey, 18th Infantry, U.S.A. Graduate of West Point, 1908.

1917-1922. Major Ralph R. Stogsdall, Q.M.D., U.S.A., retired. During the S.A.T.C., Oct. 1 to Dec. 14, military department was in charge of the

U. S. military authorities under Major E. A. Freeman. Part of the time, also—1919-1920—Major J. W. Moore, of the Citadel faculty, was Commandant,—Col. Stogsdall being P.M.S.&T.

1922-1923. Captain James C. Hutson, C.A.C., U.S.A. Graduate of The Citadel, 1913.

1923-1926. Major Albert Gallatin Goodwyn, Infantry, retired.

1926- Major Jacob A. Mack, C.A.C. Graduate of West Point, 1904. Acted as Commandant, May 1 to June 8, 1926.

1926-1931. Major William C. Miller, Infantry, U.S.A. Graduate of West Point.

1931- Lt. Colonel John W. Lang. Graduate of West Point. The military staff at this time in charge of the two units (infantry and coast artillery) consisted of seven Army officers detailed by the War Department for duty at The Citadel.

APPENDIX E

Commencement Exercises

(NOTE: During the first period of The Citadel's operation, only cadet speakers were heard at the commencement exercises.)

1846, Nov. 20th, 10:30 A.M. Second Baptist Church. (In Wentworth Street—now Centenary Church, colored.)

1847, Nov. 18th, 10:30 A.M. Military Hall, Wentworth Street (Afterwards the German Artillery Hall, which was torn down in 1930 to give space for Kerrison's automobile parkway.)

1848, Nov. 24th, 10:30 A.M. Hibernian Hall.
1849, Nov. 22nd, 10:30 A.M. Hibernian Hall.
1850, Nov. 22nd, 10:30 A.M. South Carolina Hall.
1851, Nov. 20th, 10:30 A.M. Hibernian Hall.
1852, Nov. 18th, 10:30 A.M. Hibernian Hall.
1853, No graduating class.
1854, Nov. 24th, 10:30 A.M. Hibernian Hall.
1855, Nov. 21st, 10:30 A.M. Hibernian Hall.
1856, Nov. 21st, 10:30 A.M. Hibernian Hall.
1857, Nov. 18th, 10:30 A.M. Hibernian Hall.
1858, No exercises due to change of date to April following.
1859, April 9th, 10:30 A.M. Hibernian Hall.
1860, April 11th, 10:30 A.M. Hibernian Hall.
1861, No exercises due to outbreak of war.
1862, April 8th, 10:30 A.M. Hibernian Hall.
1863, April 15th, 10:30 A.M. Hibernian Hall.
1864, April 13th, 10:30 A.M. Citadel Square Baptist Church.

The literary societies in the ante-bellum days had public exercises at commencement, the principal feature of which was an address by some prominent speaker. A list of these speakers will be found in Appendix G.

At post-bellum commencements, the custom was reversed: The annual address was delivered at the academic graduating exercises, and the literary societies—both of which kept up elaborate exercises of their own for a number of years—had only cadet speakers. (Starred dates indicate that the speaker was a Citadel alumnus.)

Speakers, Second Period

*1886, July 28th, General Ellison Capers, Hibernian Hall.
1887, July 27th, General Edward McCrady, German Artillery Hall.
1888, July 4th, Hon. Samuel Dibble, M.C., Academy of Music.
1889, July 3rd, Hon. J. J. Hemphill, M.C., Hibernian Hall.
*1890, July 2nd, Colonel John P. Thomas.
1891, July 1st, Hon. Geo. W. Dargan, M.C.
1892, July 8th, Rev. Dr. C. S. Vedder, Grand Opera House.
1893, July 14th, Hon. Jos. W. Barnwell, Esq., Courthouse, Aiken, S. C.
1894, June 29th, Hon. John D. Kennedy, Courthouse, York, S. C.
1895, June 28th, Hon. Leroy F. Youmans, Camden, S. C.
1896, June 30th, Hon. John L. McLaurin, U. S. Senator (S.C.), Sumter, S. C.
1897, June 30th, Judge Joshua H. Hudson.
*1898, June 28th, Hon. Robert Aldrich.
1899, June 30th, Hon. Benj. R. Tillman, U. S. Senator (S. C.), Orangeburg, S. C.
1900, June 27th, Hon. Stanyarne Wilson, M.C.
1901, June 28th, Hon. R. B. Scarborough, M.C.
1902, June 30th, Rev. Jas. A. B. Scherer, D.D.
1903, June 30th, General Edward McCrady, Winthrop College, Rock Hill, S. C.
1904, June 30th, Hon. George S. Legare, M.C.
1905, June 30th, Professor Wm. Spencer Currell, Opera House, Columbia, S. C.
*1906, June 29th, Hon. M. L. Smith.
1907, June 29th, Governor M. F. Ansel, Auditorium, Jamestown Exposition, Norfolk, Va.
*1908, June 29th, Professor Wm. E. Mikell.
*1909, June 30th, Colonel Asbury Coward, The Pavilion, Isle of Palms.
*1910, June 14th, Captain Ellison A. Smyth, Waller Hall, Lander College, Greenwood, S. C.
*1911, June 28th, Dr. James P. Kinard, Academy of Music.

1912, June 13th, Professor A. W. Goodspeed, Academy of Music.
1913, June 13th, Hon. Fitzhugh McMaster, Academy of Music.
1914, June 15th, Professor H. S. McGillivray, Ph.D., Academy of Music.
1915, June 15th, President H. N. Snyder.
1916, June 15th, Dr. Wm. T. Ellis.
1917, May 31st, Dr. J. G. de Roulhac Hamilton.
1918, May 24th, Major William Cain, Hibernian Hall.
*1919, June 13th, Mr. William S. Lee, Hibernian Hall.
1920, June 9th, Dr. C. Alphonso Smith.
*1921, June 14th, Colonel W. W. Lewis.
*1922, June 13th, Colonel James G. Padgett.
1923, June 12th, Major General David C. Shanks, Citadel Chapel.
1924, June 10th, Dr. G. L. Swiggett, Citadel Chapel.
*1925, June 9th, Dr. T. P. Harrison, Alumni Hall, The Citadel.
1926, June 8th, Major General Johnson Hagood, Alumni Hall.
1927, June 7th, General Lawrence D. Tyson, U. S. Senator (Tenn.), Alumni Hall.
*1928, June 5th, Mr. W. E. Woodward, Alumni Hall.
1929, June 4th, Professor Archibald Rutledge, Alumni Hall.
1930, June 3rd, Major General Hanson E. Ely, Alumni Hall.
1931, June 5th, Professor Jas. Southall Wilson, Alumni Hall.
1932, June 3rd, Colonel Robert R. McCormick, Alumni Hall.

APPENDIX F

Baccalaureate Sermons

(NOTE: Any information which will enable the Dean of The Citadel to fill in the blank spaces below will be received with thanks.)
1845.
1847.
1848.
1849.
1850.
1851, Rev. Christian Hanckel, D.D., St. Paul's Church.
1852.
1853.
1854.
1855, Rev. J. R. Kendrick, Citadel Square Baptist Church.
1856.
1857.
1858, Commencement changed to following spring.

1859, Rev. J. L. Girardeau, Citadel Square Baptist Church.
1860.
1861.
1862, Rev. E. T. Winkler, First Baptist Church.
1863, Rev. B. M. Palmer, "Dr. Smyth's Church" (Second Presbyterian).
1864, Rev. J. R. Pickett, Second Presbyterian Church.

*1886, Rev. O. A. Darby, D.D. (Methodist).
1887, Rev. Charles Manley, D.D. (Baptist).
1888, Rev. C. C. Pinckney, D.D. (Episcopalian).
1889, Rev. J. Wm. Flynn, D.D. (Presbyterian).
1890, Rev. R. D. Smart, D.D. (Methodist).
1891, Rev. John Gass (Episcopalian).
1892, Rev. E. C. Dargan, D.D. (Baptist)
1893, Rev. W. M. Grier (A. R. Presbyterian).
1894, Rev. W. R. Atkinson, D.D. (Presbyterian).
1895, Rev. Ellison Capers, D.D. (Episcopalian).
1896, Rev. Lucius Cuthbert (Baptist).
1897, Rev. J. A. Clifton (Methodist).
*1898, Rt. Rev. P. F. Stevens (Episcopalian).
1899, Rev. H. W. Bays, D.D. (Methodist).
*1900, Rev. Geo. H. Cornelson (Presbyterian).
*1901, Rev. Albert S. Thomas (Episcopalian).
*1902, Rev. John G. Beckwith (Methodist).
*1903, Rev. F. W. Gregg (Presbyterian).
*1904, Rev. W. P. Witsell (Episcopalian).
*1905, Rev. A. N. Brunson (Methodist).
*1906, Rev. S. H. Booth (Methodist).
1907, Rt. Rev. A. M. Randolph (Episcopalian), St. Paul's Church, Norfolk, Va.
*1908, Rev. Geo. H. Atkinson (Presbyterian).
*1909, Rev. B. G. Murphy (Methodist).
*1910, Rev. Jno. O. Willson, D.D. (Methodist).
*1911, Rev. T. M. Hunter (Presbyterian).
*1912, Rev. John Kershaw (Episcopalian).
*1913, Rev. W. C. Alexander (Presbyterian).
*1914, Rev. R. E. Gribben (Episcopalian).
*1915, Rev. H. H. Thayer (Baptist).
*1916, Rev. J. C. Bailey (Presbyterian).
*1917, Rev. J. H. Noland (Methodist).

1918, Rev. Mercer P. Logan, D.D. (Episcopalian).
*1919, Rev. S. C. Morris (Methodist).
*1920, Rev. C. C. Derrick (Methodist).
1921, Rev. William Way, D.D. (Episcopalian).
1922, Rev. C. C. Coleman (Baptist).
*1923, Rev. John Lake (Baptist).
1924, Rev. Geo. J. Gongaware (Lutheran).
1925, Rt. Rev. Wm. A. Guerry, D.D. (Episcopalian).
1926, Rev. Robert G. Lee (Baptist).
1927, Rev. J. W. Hickman, D.D. (Presbyterian).
1928, Rev. Carl S. Smith (Episcopalian).
1929, Rev. Walter B. Capers, D.D. (Episcopalian).
1930, Rev. D. I. Purser (Baptist).
1931, Rev. Chas. B. Foelsch (Lutheran).
*1932, Rev. Eben Taylor (Methodist).

APPENDIX G
Ante-Bellum Literary Society Commencement Speakers

In the early days of the Citadel, only cadets made addresses at graduating exercises, which were held in the forenoon. The literary societies had their exercises in the evening—of some day in "commencement week." The principal feature of these occasions was the annual address, made by some prominent speaker, although there were cadet "salutatorians" and "valedictorians."

After the War, the custom was reversed—the annual address being delivered at the graduating exercises, and the societies having cadet speakers.

Nov., 1846, The Calliopean Society was addressed by its founder, Professor F. W. Capers, of the Citadel faculty. (The Polytechnic Society was founded in 1847. After the formation of the Polytechnic Society, the annual address was made before the two societies jointly.)
Nov., 1847, Hon. S. W. Trott, in Temperance Hall.
Nov., 1848, B. F. Hunt, Esquire, in Hibernian Hall.
Nov., 1849, Colonel James Simons, in Hibernian Hall.
Nov., 1850, Colonel Samuel McGowan, in Carolina Hall.
Nov., 1851.
Nov., 1852, Colonel L. M. Keitt.
Nov., 1853, Richard Yeadon, Esquire. There were no graduating exercises this year owing to an émeute of the Second Class the year before, and Mr. Yeadon took for his subject the timely topic, "The Necessity of Subordination in Our Colleges."

Nov., 1854, Hon. Daniel Wallace.
Nov., 1855, Wm. H. Trescott, Esquire, in Hibernian Hall.
Nov., 1856, Captain J. D. Tradewell, in Hibernian Hall.
Nov., 1857, Geo. D. Bryan, Esquire, in Military Hall.
Nov., 1858. No exercises due to change of date of commencement.
April, 1859, B. R. Carroll.
April, 1860, Hon. W. D. Porter.
April, 1861. No exercises on account of outbreak of the war.
April, 1862, Colonel A. P. Aldrich, in Hibernian Hall.
April, 1863, Judge A. G. Magrath, in Hibernian Hall.
April, 1864, Rev. P. F. Stevens, in "Rev. Mr. Girardeau's Church."

APPENDIX H
Best-Drilled Cadet in the Corps—"Star of the West" Medal

(NOTE: The first contest to determine the best-drilled cadet in the corps was held in 1886. It became an annual event thereafter in connection with the commencement exercises. The *Star of the West* Medal was first awarded to the winner of the title, "Best-Drilled Cadet" in 1893.)

1886 J. T. Coleman
1887 W. C. Davis
1888 J. R. Rutledge
1889 P. K. McCully
1890 W. Z. McGhee
1891 A. S. Thomas
1892 J. S. Verdier

Winners of the "Star of the West" Medal

1893 A. E. Legare
1894 A. Levy
1895 J. D. Dial
1896 J. M. Josey
1897 J. B. Salley
1898 D. C. Pate
1899 L. B. Steele
1900 A. H. Cross
1901 T. C. Marshall
1902 A. E. Hutchison
1903 J. F. O'Mara
1904 E. C. Register
1905 W. W. Dick
1906 W. W. Benson

1907 A. T. Corcoran
1908 E. D. Smith
1909 D. W. Gaston
1910 F. Y. Legare
1911 Thos. Richardson
1912 J. M. Arthur
1913 J. H. Holmes
1914 James Anderson
1915 J. G. M. Nichols
1916 F. R. Rogers
1917 H. L. Cunningham
1918 T. W. Williamson
1919 J. L. Whitten
1920 E. A. Pollock
1921 J. D. Frost, Jr.
1922 E. T. Moore
1923 Walter Allan
1924 J. J. Mackay
1925 C. H. Rossen
1926 F. G. Burnett
1927 E. B. Fishburne
1928 W. M. Roberts
1929 R. K. Walker
1930 J. W. Blevins
1931 R. A. Zobel
1932 R. H. Ammerman

APPENDIX I
"Oldest Living Graduate"

No.	Class	Name	Period Holding Title	No. Years
1.	1846	C. C. Tew	Nov. 20, 1846, to Sept. 17, 1862	16
2.	1846	R. G. White	Sept. 17, 1862, to May 20, 1875	13
3.	1846	C. O. Lamotte	May 20, 1875, to March 25, 1883	8
4.	1846	J. L. Branch	March 25, 1883, to Jan. 15, 1894	11
5.	1847	Johnson Hagood	Jan. 15, 1894, to Jan. 4, 1898	4
6.	1847	E. L. Heriot	Jan. 4, 1898, to Dec. 6, 1903	5
7.	1849	P. F. Stevens	Dec. 6, 1903, to Jan. 9, 1910	7
8.	1850	J. A. Houser	Jan. 9, 1910, to March 15, 1910	2 mos.
9.	1850	J. W. Robertson	March 15, 1910, to March 5, 1911	1
10.	1851	J. P. Thomas	March 5, 1911, to Feb. 11, 1912	1
11.	1851	W.D. McMillan	Feb. 11, 1912, to Dec. 25, 1913	2

No.	Class	Name	Period Holding Title	No. Years
12.	1852	C. S. Gadsden	Dec. 25, 1913, to Jan. 11, 1915	1
13.	1854	Asbury Coward	Jan. 11, 1915, to April 28, 1925	10
14.	1861	C. I. Walker	April 28, 1925, to Nov. 7, 1927	2
15.	1861	R. O. Sams	Nov. 7, 1927, to March 4, 1930	3
16.	1886	T. P. Harrison	March 4, 1930—	

APPENDIX J

The faculty in 1882 and in 1932

Academic Board (Faculty) 1882

Colonel J. P. Thomas, Superintendent and Professor, History, Belles Lettres, and Ethics.
Major William Cain, Professor, Mathematics and Engineering.
Major DelKemper, Professor, Chemistry and Physics.
Captain Lyman Hall, Assistant Professor, Mathematics, and in charge of Drawing.
First Lieutenant P. P. Mazyck, Assistant Professor, in charge of French.
First Lieutenant H. T. Thompson, Assistant Professor, History and Belles Lettres.
Second Lieutenant E. M. Weaver, Junior, 2d Artillery, U.S.A., Professor of Military Science and Tactics.
F. L. Parker, M.D., Surgeon.
Second Lieutenant W. W. White, Quartermaster.

Faculty 1932
(In Order of Seniority)

General Charles P. Summerall, LL.D., President.
Colonel Oliver James Bond, B.S., LL.D., Professor of Mathematics.
Lieutenant Colonel John Walton Lang, Infantry, U.S.A., Professor of Military Science and Tactics.
Major Hugh Swinton McGillivray, A.B., Ph.D., Professor of English.
Major Louis Knox, B.S., M.S., Professor of Chemistry and Biology.
Major Louis Shepherd LeTellier, M.S., Professor of Civil Engineering.
Major Newland Farnsworth Smith, Ph.D., Professor of Physics.
Major Smith Johns Williams, A.B., A.M., Professor of History.
Major Harold Carter Winship, A.B., A.M., Professor of Modern Languages.
Major Eugene Villaret, Coast Artillery, U.S.A., Associate Professor of Military Science and Tactics.
Major Clifton LeCroy Hair, B.S., Professor of Mathematics.
Major John Anderson, C.E., Professor of Engineering and Drawing.

Major Leonard Augustus Prouty, A.B., Professor of Psychology and Education.
Major Marion Smith Lewis, A.B., M.A., Professor of Business Administration.
Major Paul Lewis Ransom, Infantry, U.S.A., Associate Professor of Military Science and Tactics.
Captain Carl Francis Myers, Jr., B.S., Associate Professor of Mathematics.
Captain Lewis Simons, Infantry, U.S.A., Assistant Professor of Military Science and Tactics.
Captain Alfred E. Dufour, A.B., Associate Professor of Modern Languages.
Captain Milton Boone Kennedy, A.B., A.M., Associate Professor of English.
Captain James Karl Coleman, B.S., A.M., Associate Professor of History.
Captain Alston Deas, Infantry, U.S.A., Assistant Professor of Military Science and Tactics.
Captain William Q. Jeffords, Junior, Coast Artillery, U.S.A., Assistant Professor of Military Science and Tactics.
Captain Ralph Milledge Byrd, B.S., M.S., Ph.D., Associate Professor of Chemistry.
Lieutenant Ralph Muse Lyon, B.S., A.M., Assistant Professor of Education.
Lieutenant Luther Brenner, B.S., M.S., Assistant Professor of Physics.
Lieutenant St. Julien Ravenel Childs, A.B., M.A., Assistant Professor of History.
Lieutenant John Alvah Lee Saunders, B. S., M.A., Assistant Professor of Mathematics.
Lieutenant Charles T. Razor, B.S., Assistant Professor of Physics.
Lieutenant J. Alvin Tiedemann, B.S., Assistant Professor of Business Administration.
*Lieutenant James Geraty Harrison, B.S., B.Litt., Assistant Professor of English.
Lieutenant Joe Henry Watkins, B.S., Assistant Professor of Chemistry.
Lieutenant Ephraim Clark Seabrook, B.S., Assistant Professor of Mathematics.
Lieutenant Granville Paul Smith, Assistant Professor of History.
Lieutenant Hugh Dudley Ussery, M.A., Assistant Professor of Physics.
Lieutenant Hilliard Galbraith Haynes, A.B., B.S. in C.E., Assistant Professor of Engineering.
Lieutenant William Sylvester Price, A.B., Assistant Professor of Modern Languages.

---

*On leave of absence for one year.

Lieutenant Francis Burt Fitch, Junior, C.E., Assistant Professor of Engineering.
Lieutenant Larry Jordan Willis, A.B., M.A., Assistant Professor of Modern Languages.
Lieutenant Frank Cambridge Tibbetts, B.S., M.B.A., Assistant Professor of Business Administration.
Lieutenant Ray E. Dingeman, Coast Artillery, U.S.A., Assistant Professor of Military Science and Tactics.
Lieutenant Robert Waller Achurch, A.B., M.A., Assistant Professor of English.
Lieutenant Paul Rupard Sanders, A.B., Assistant Professor of English.
Lieutenant John Andrew Hamilton, Junior, A.B., M.A., Assistant Professor of Modern Languages.

INDEX

INDEX

A.E.F. Scholarships, 220.
Aldrich, Lieutenant Alfred, 65; account of Cadet Company, 66-71; 188.
Anderson, Major Robert, 48, 49.
Annual encampment, experiment (1854), 43; first (1889), 124; changed to march, 162-164; second annual march, 165; rifle practice (1911), 175.
Arsenal, 14, 17-19, 41-42; guard, act of 1822, establishing, 4; cadets, 43, 55, 60, 64, 65, 74, 82-85; academy, 41-42, 81, 82, 169.
Association of Graduates, organization of, 45; orators at annual meeting, 46; meets to reopen school, 91-92; reorganization and personnel of (1887), 95; meeting for reopening school, 95-98; Roll of Honor, 96; officers of (1877), 98; campaign to reopen school, 99, 102; advocating Citadel, 125; toasts to, at Semicentennial, 131, 133, 149; publishes *Bulletin*, 156; meeting of (1903), 159; discusses changing name of school from *academy* to *college*, 167-170; donates funds for rifle team, 182; reunion of (1920), 196; officers of, See Appendix C.

Band, organized, 171.
Battalion of State Cadets, The, act establishing (1861), 55-56, 63; at battle of Tulifinny, 74-78; 82; 93.
Beneficiary Cadets, provisions for, 24, 34, 57; sixty-eight scholarships established (1882), 107; compared with pay cadets, 111; character of, 128; 220-221.
Blue Book, first compiled, 164.
Board of Visitors, organization and first personnel of, 17-22; 24, 25; report on dismissal, 27; yellow fever resolutions, 35-37, 38; expulsion of Second Class (1852), 39, 40; proposed consolidation with arsenal, 41-42; discipline, 42-43; no commencement, 51-52; personnel of (1861), 52; report on enrollment (1862), 55; financial report, 56-57; refuses unconditional return of buildings, 94; meeting to reopen school, 97; at reopening, 103, 107; abolishes fraternities, 113; dismisses sixty-four cadets (1898), 146-148; financial report, 150-151; granting of right to confer degrees, 151-152; regulations for appointment to, 152; approves elective system, 156-157; awards to Coward and Simons, 166; changes name of school, 170; 171-172, 174; changes site, 194; provisions for new buildings, 194; raises academic standards, 203; report to Superintendent of Education, 209-212; report, 214, 215; awards degrees, 215-216; chairmen of, See Appendix A.
Bond, Colonel Oliver James, xi, 120; appointed Superintendent, 166; report in *Bulletin*, 166-167; suggestion to change name of school, 167-170; report of (1911), 176; telegram to Colonel Leonard Wood, 183; at Winthrop, 184; 196; becomes Dean, 219.
Branch, Colonel J. L., 23, 32, 46, 49.
Brisbane, Colonel Abbott H., 21, 40.
Buildings, first Citadel building ("Tobacco Inspection"), 1, 4, 5, 6, 7, 12; description of (1870), 88; description of, by Federal officer, 89-90; fire of 1869, 90; efforts to recover from United States, 92-95; buildings evacuated by Federals, 94; damage by earthquake, 122-123; seal placed on building, 123; fire of 1892, 125-126; restoration, 126-127; improvements, 160; city guard-house obtained, 162; gymnasium constructed, 162; new buildings, 171-172, 173-175; plans for greater Citadel approved, 196-197; progress of building, 197; corner stone of greater Citadel laid, 198; construction, 199-202; Murray Barracks built, 209; new college building, 210; homes

237

238 INDEX

for officers, 210-211; use of Old Citadel, 211-212; movement for construction of main college building, 212-215.

Bulletin, first published (1903), 156; resolutions on curriculum, 156-157; Colonel Coward's retirement, 165; Major Bond's report, 166-167; advocates changing name of school, 167-170; Colonel Coward's speech, 170; 172, 173, 174.

Burgess Mill, Battle of, 69.

Cadet Company, The, footnote on, 64; account of, 65-71.

Cadet life, regulations (1849), 28; Fourth of July Association, 30; room equipment and duties, 107-109; leave and diversions, 111-112; extracurricular activities, 112; fraternities, 112, 113; Blue Book, 164.

Cain, Major William, 113-114, 188.

Calhoun, John C., 12, 13, 26; death of, 34-35; 129, 153.

Calliopean Society, 29; account of, by C. I. Walker, 30; 32; reorganized, 112.

Capers, Colonel Ellison, officer of rifle regiment, 49; resignation and war service in West, 52-53; at Semicentennial, 131.

Capers, Major F. W., 18, 29n; receives W.L.I. colors, 44; 79, 103.

Chamberlain, Governor Daniel, 90-91, 92, 93, 94.

Charleston, 6; cyclone (1885), 117-118; earthquake (1886), 121, 123; West Indian Exposition, 155.

Citadel, The, origin of, 1, 16, 17, 19; aim of, 19; organization of, 17; curriculum, 25, 28; discipline, 41, 42-43; enrollment (1862), 54, 55; property destroyed by Sherman, 80; campaign to reopen after the war, 91, 104; capacity (1882), 109; enrollment (1882), 110; difference from original plan, 110; curriculum (1883), 113-115; decreased enrollment (1886), 119; enrollment (1892), 128; military training (1894), 137; trips taken, 141, 142; enrollment (1899), 150; conferring of degrees, 151-152; enrollment (1902-1912), 156; elective curriculum, 157; growth (1905), 162; condition of (1908), 166-167; name changed, 167-170; condition in 1909, 171; scholastic standards, 176; construction of Greater Citadel, 198-202; raising academic standard, 203-208; Business Administration Department established, 207; elective system, 207, 208; enrollment (1922-1932), 208, 209; C.E. degree established, 215.

Citadel News, description of greater Citadel, 199-201; on progress of building, 197; February, 1920, 196-197.

Clemson College, 119, 125, 129, 145.

Coffin, George M., account of Battle of Tulifinny Creek, 76-78.

Colcock, Major R. W., 23, 32, 35, 39; resignation of, 40; 79.

Coleman, James Thomas, best drilled cadet in U. S., 116.

College Regulations (1849), 28-29; (1898), 146-147; (1906), 164.

Commandant of Cadets, office created, 124; McMaster relieved, Simons installed, 161; list of P. M. S. & T's, See Appendix D.

Commencements, 23, 32, 33; time of changed, 38; procedure at the first, 31; dispensed with (1861), 51-52; date of changed, 176; last at old Citadel, 202.

Confederacy, 22; organization of, 51.

Confederate Monument (Magnolia Cemetery), unveiling of, 110.

Congress of U. S., discussion on tariff, 9; investigates commissioning of Citadel men in Marine Corps, 185.

Coward, Colonel Asbury, 30, 32, 39; at meeting of graduates, 46; 100; elected Superintendent, 124; 126, 131; takes Corps to Aiken, 136; 138, 139, 143, 148; Superintendent, 160,

INDEX 239

161; *Bulletin* on, 165; presented cup, 166; address of 1869 and 1909, 170-171; at Diamond Jubilee, 185.
Cumming's Point, construction of batteries at, 49.

David, Lieutenant John H., 186, 187, 190.
Dean, Colonel John Mills, 59.
Degrees, right to confer, 151-152; academic, 215; honorary, 215-216; holders of honorary degrees, 216.
Diamond Jubilee, exercises, 185.

Faculty, First, 18; homes for, 210-211, 212; of 1882 and 1932, See Appendix J.
Financial statement (1864), 57.
Finley, Dr. J. E. B., 18, 22, 23.
Fire, of 1869, 90; of 1892, 125, 126.
Flags, presented by Washington Light Infantry, 44; 83, 84, 85.
Fraternities, evils of, 112; abolished, 113.
Furman University, 54; first annual encampment at, 123-124; 206; football game with, 217,219.

Gadsden, Major C. S., 92, 98, 100, 122, 145.
General Assembly, 4, 10, 19, 21, 25, 87; bill for reopening the school, 100-103; visits Citadel (1895), 140-141; law governing degrees, 151; regulates appointment of Board of Visitors, 152; appropriates funds for building improvements, 160; visits Citadel, 171-172; (1917), 183; makes appropriations for Greater Citadel, 194, 196-197.
Graham, Captain W. F., 18, 22-23, 43. See Appendix B.
Grange Movement, its origin in the South, 118; rise of Benjamin R. Tillman, 119.
Greater Citadel, need and move for, 192; choice of site, 193; appropriation for, 194; plans for approved, 196-197; progress of building, 197;
laying of corner stone, 198; description of, 199; first session in, 206; 207, 210.

Hagood, General Johnson, 32n, 46; on General Micah Jenkins, 61; on repossessing buildings, 94; 98, 103; speech at semicentennial, 129-130; response to toast at semicentennial, 131-132; death of, 145.
Hamilton, James, Jr., 2, 3.
Hammond, Governor J. H., 9, 14, 17, 20.
Hammond, Senator James H., 215.
Hampton, General Wade, 67, 90, 91, 92, 94, 98.
Haynesworth, Cadet G. E., first shot of War fired by, 50.
History of South Carolina Military Academy, v, 20n, 26, 34n, 58, 61-62, 63, 78, 83-84.
Holmes, Captain James H., 189, 191.
Homecoming, first (1924), 216-219.
Hume, Dr. William, 23, 37, 40, 78, 80.
Humphrey, Captain M. B., 66-67, 68, 70.

Jamison, General D. F., 17, 21, 38n; speaker at meeting of Association of Graduates, 46; President of Secession Convention, 47, 52, 103.
Jenkins, Lieutenant John M., work at Citadel, 135; commendation of, 137-138; tribute of Board, 143-144.
Jenkins, General Micah, 59; death of, 61-62; 135, 138.
Johnston, General G. D., Superintendent, 118, 159; resigns, 124; See Appendix B.
Jones, General James, 16, 18, 21, 52, 63, 64, 102; See Appendix A.
Jones, General Sam, 33n, 65, 74, 98.

Kennedy, General John D., casts deciding vote to reopen school, 102, 103; commencement address, 139.
Kinard, Dr. J. P., 120, 121, 176, 184, 203, 216.

Last shot of War fired, 85.

240 INDEX

Law, Colonel E. M., 59, 60.
Law, T. H., 98, 133.
Lee, William S., at Catawba project, 157, 177; W. S. Lee Scholarship, 197, 221; plans for athletics, 196, 216.
Library, Sherman destroys, 80; beginning of (1888), 125; need of new facilities, 214.
Little, A. H., in Mexican War, 34.
Louisa Courthouse, Battle of, 67.

Magazine, 4-5, 11.
McCants, E. C., 205-206, 216.
McCrady, General Edward, 157, 158.
McDuffie, Governor George, 12, 13, 14.
Means, General John H., 17, 21, 22, 46, 52; death of, 60, 103.
Mercury, The Charleston, 7, 50, 51,
Military Education, purpose of, 178; training, 180; report on (1915), 180-181; Colonel Bond, on, 182.
Mitchell, W. H. J., The Story of a Forgotten Hero, 71-74.
Mood, Captain Julius A., 189, 191.
Morris Island, fortified to protect harbor, 49; and *Star of the West,* 50, 64.
Moses, J. H., 69-70.
Fort Moultrie, 7, 10; Federal troops occupy, 48; Anderson evacuates, 49; 171; used for training cadets, 180.
Murray, A. B., 209, 211.

National Defense Act, 182.
National Rifle Match, participated in (1916), 182.
Negroes, slave uprising, 1-4; breach between races, following War, 86; gain entrée to white schools, 87; trouble with (1876), 90; burial of negro fifer, 137.
Nettles, Lieutenant W. A., 69.
New Orleans Exposition, trip to, 115-116.
Nullification, 9, 11-13, 21.

Old Guard, the, picture made of, 159.

Ordnance stores, Citadel first Confederate laboratory to manufacture for, 52.

Padgett, Colonel James G., 194, 196, 202.
Palmetto Regiment, 33-34.
Palmer, R. A., 58.
Parker, Captain Charles, 11, 16.
Partridge, Captain Alden, 14n, 18.
Pickens, Governor F. W., 48, 49, 56, 193.
Polytechnic Society, 29; account of by C. I. Walker, 30; 32; reorganized, 112.

Reconstruction, 85-87, 90-91.
Richardson, Governor J. P., 15, 16, 17, 102.
Roll of Honor, War Between the States, 96; World War, 190.
R. O. T. C., value of, 179; 190.

Sams, R. O., 75, 216.
Scholarships, sixty-eight state scholarships (1882), 107, 111, 128; City of Charleston scholarships, 145; 205; A.E.F. scholarships (1932), 220; Crouch scholarship, 221; W. S. Lee scholarship, 221.
Secession, 21, 22; convention to consider, 47; signing of Ordinance of, 47-48.
Semicentennial, The, 129-135.
Shell Trophy in Bond Hall, 80-81.
Sheppard, Orlando, 198, 202, 216.
Simons, Captain W. H., becomes commandant, 161; takes corps on march, 161-164; revises regulations, 164; compiles records, 164; presented saber, 166.
Sims, Major R. M., 83-84.
Slave Uprising, 1-4.
Solar Eclipse, cadets view, 153-154.
South Carolina, 9; division of sections, 140.
South Carolina College, 21; as hospital (1862), 54; opened to negroes,

INDEX 241

87; 103, 125, 152; football game with, 198-199; 206.

South Carolina Volunteers, 14th, 21.

Southern Association of Colleges, resolution to seek membership in, 203; Citadel admitted into, 206.

Spanish-American War, 145; strength of Spanish Army and Navy, 146; Fort Sumter armed, 146; Declaration of War, 148; South Carolina's contribution, 148, 149; Citadel's contribution, 148, 149.

Star of the West, Citadel cadets fire upon, 49-51; 56, 63, 94; medal founded, 136; first medal competition and winner, 136; winners of medal, See Appendix H.

Stevens, Major P. F., Secretary of Association of Graduates, 46; speaker at meeting of graduates, 46; Commander of Citadel cadets, 49; 50; cited for *Star of the West* service, 52; service in Virginia, 52, 60; 92, 96, 98, 100; becomes oldest alumnus, 159; superintendent, 159.

Students' Army Training Corps, established, 189; conditions under, 190.

Summerall, General Charles P., reviews corps (1927), 209; accepts Citadel presidency (1931), 219; record, 219-220.

Sumter, Fort, 46, 48; Major Anderson transfers to, 49; 52, 56, 63; fortified for war of 1898, 146.

Superintendents, history of, 159; list of, See Appendix B.

Teague, Dr. B. H., founds *Star of the West* medal award, 136.

Tew, Colonel Charles Courtenay, 19; proto-graduate, 31; 33n; president of Association of Graduates, 45; 46; killed at Sharpsburg, 60.

Thomas, Bishop A. S., 155.

Thomas, Colonel J. Pulaski, 194, 196, 202; report to Board of Education, 209-212; report to legislature (1928), 212-213; report (1929), 213-214; report (1930), 214-215.

Thomas, Colonel J. Peyre, v, 22, 26, 39, 40; Secretary of Association of Graduates, 45; speaker at meeting of graduates, 46; cited for *Star of the West* service, 52; 74, 75, 78, 81, 82, 83, 84, 85; poem on the deserted Citadel, 88; sketch of the academy, 98; 100, 102, 145; Superintendent of Citadel, 110-111; Professor of English, 114; resigns as Superintendent, 116-117; 118, 131; semicentennial telegram, 135; 143, 145, 148, 160; death of, 177; *Sketch of S. C. Military Academy,* 178; 186.

Thompson, Governor H. S., 63, 65, 75, 98, 100.

Tillman, Benjamin, rises to prominence, 119, 126; establishment of Clemson and Winthrop, 129; at semicentennial, 130; 132; partiality for Clemson, 125; speaker at Commencement, 150.

"Tobacco Inspection," 1, 4, 5, 6, 7.

Trevilians, Battle of, 67-68.

Trips taken by Cadets, New Orleans, 115-116; Furman, 124; Aiken, 136; Rock Hill, 138, 157; Camden, 141; Atlanta, 142; Savannah, 142; Sumter, 142, 143; Anderson, 145; Barnwell, 145; Orangeburg, 150; Wadesboro, 153; Florence, 154; Darlington, 154; Winthrop, 158; St. Louis, 160; Norfolk, 164; Greenwood, 172-173.

Tulifinny, Battle of, 65, 74-78.

U. S. Army, 7, 8, 10; account of occupation of Citadel by, 91-94; evacuates buildings, 94.

Vesey, Denmark, 1, 2, 3.

V. M. I., 15, 16, 18, 148, 155, 188, 189, 219-220.

Walker, General C. I., 30; honor graduate (1861), 53; war service, 53; 92, 98, 100, 104.

Walker, Robert Murdoch, 121; contractor of New Citadel, 194-195; editorial on, 195, 196.

Wallace, Dr. D. D., 14n.

War Between the States, outbreak, 46; cadets at Cumming's Point, 49; first shot, 50; graduates who served as Confederate officers, 58; officers' records in war, 58-62; cadets' war record valid, 62, 63; service around Charleston, 63-65; Cadet Company, 64-71; Battle of Tulifinny, 74-78; fall of Charleston, 78-82; last days of war, 82-85; last shot fired, 85.

Washington Light Infantry, 7, 8, 23n, 35; Citadel presented colors by, 44; Semicentennial of, 44; armory used as meeting place, 95; 97, 98, 99; celebration (1879) dedicated to Citadel, 98; Washington Day celebration (1883), 110; at Semicentennial, 130; toast at Semicentennial, 133; speech on relation to Citadel, 134.

Weaver, E. M., 114-115; commander of New Orleans Platoon, 115; acting Superintendent (1885), 117.

Weeks, Teddy, 218.

Welch, S. E., description of *Star of the West* incident, 49.

Wesner, Frederick, 7.

West Indian Exposition, 155.

West Point, 17, 18, 22, 40-41, 137, 167-169, 204.

White, Major J. B., 63, 64; account of Battle of Tulifinny, 74-76; 78, 98, 152; Superintendent, 159; See Appendix B.

Willson, Dr. John O., baccalaureate sermon, 172; establishes "John O. Willson Ring" award, 173; telegram from, at Diamond Jubilee, 185.

Willson Ring, founded, 173.

Winthrop, 152, 157, 158, 184, 203.

Woodward, W. E., 216.

World War, Citadel volunteers services, 183; graduates commissioned (1917), 184, (1918), 188; war record reviewed, 186-188; training at Plattsburg, N. Y., 189-190; Students' Army Training Corps, 189, 190; Armistice, 190; number of graduates in service, 190; Roll of Honor, 190-192.

Yellow fever, 35; resolutions of Board concerning, 35-38; origin and effects of, 36-38; transfer of cadets to Arsenal Academy, 38; in 1871, 90.

Y. M. C. A. founded (1886), 120.

Youmans, Leroy F., 141.

Zouave Cadets, 49; in *Star of the West* incident, 50.

www.ingramcontent.com/pod-product-compliance
Lightning Source LLC
Chambersburg PA
CBHW031410290426
44110CB00011B/332